MOOD MATTERS

From Rising Skirt Lengths
to the Collapse of World Powers

John L. Casti

MOOD MATTERS

*From Rising Skirt Lengths
to the Collapse of World Powers*

Copernicus Books

An Imprint of Springer Science + Business Media

Springer Science+Business Media, LLC © 2010

All rights reserved. No part of this publication may be reproduced, stored in a retrieval system, or transmitted, in any form or by any means, electronic, mechanical, photocopying, recording, or otherwise, without the prior written permission of the publisher.

Published in the United States by Copernicus Books,
an imprint of Springer Science+Business Media.

Copernicus Books
Springer Science+Business Media
233 Spring Street
New York, NY 10013
www.springer.com

Library of Congress Control Number: 2010922994

Manufactured in the United States of America.
Printed on acid-free paper

ISBN 978-3-642-04834-0   e-ISBN 978-3-642-04835-7

To Robert R. Prechter, Jr.—from a follower afar

Further Advance Praise for Mood Matters

"Homo economicus is extinct, replaced by Homo socionomicus. People are not rationally calculating utility maximizing machines. We are a social primate species driven by emotions and herd instincts, and the "animal spirits" that push and direct the economy come to life in this ground-breaking work that promises to shake up how we all view the economy in this post meltdown world. Economists may disagree with Casti on his heavy emphasis on the social mood of society, but they ignore him at their peril."

Michael Shermer, *Publisher of Skeptic magazine, columnist for Scientific American, author of "The Mind of the Market", and professor at Claremont Graduate University.*

"They [the chapters] tell an engrossing story, and the mystery heightens as it goes. . . . it's chatty and knowing."

Greg Benford, *Physicist and science-fiction writer, author of "Timescape" and "Deep Time"*

"This is an unusual yet important book. If many historians think history is just 'one damned thing after another,' driven by chance, John Casti argues that human emotion and belief drive history and endow it with foreseeable patterns. The 'socionomics' he champions may seem peculiar in the current intellectual climate, but its ideas will likely be part of a future and more scientific understanding of human events"

Mark Buchanan, *Author of "The Social Atom" and "Ubiquity"*

"I am struck by how thought-provoking it all is. I am sure that this book will draw a lot of attention and create debate."

Tor Nørretranders, Science writer, author of "The Generous Man" and "The User Illusion"

Preface

I n the pages that follow, readers will be asked to suspend belief in one of the most widely-held ideas about how the world works. Conventional wisdom has it that collective social events—the outbreak of war, the popularity of a certain film, a shift in political ideology—cause a population to feel a certain way about their future. Here we argue just the opposite: The way a population feels is what shapes events. Moreover, there is no feedback from those events to the overall social mood. In short, events don't matter, at least insofar as what causes a group to feel one way and not another about its prospects for the future. This is the distilled essence of the case made here for understanding the way the world *really* works. Now let me say a word or two about how and why I came to write this book.

About 20 years ago, I was writing a book on prediction and explanation in science. One of the chapters in that volume was devoted to the forecasting of stock market price movements. As background for that chapter, one of the people I contacted was legendary analyst Robert Prechter, who is most well known for his work on the so-called Elliott wave principle for modeling the movement of stock prices.

Over the years I stayed in contact with Prechter, and discovered that as an outgrowth of his work on the Elliott wave principle he was developing a theory of how the beliefs a population holds about the future biases the types of collective social events that were more or less likely to occur. He called this line of investigation "socionomics" and summarized the argument in a pioneering book, *The Wave Principle of Human Behavior and the New Science of Socionomics*, published in 1999.

I followed closely the subsequent development of socionomics by Prechter and his colleagues at Elliott Wave International and the Socionomics Institute, and was continually amazed at how their line of reasoning as to why events unfold the way they do and not some other way. Their arguments provided what to my eye was a much stronger and more convincing approach to understanding human affairs than the conventional wisdom noted above.

Now that I've spent a decade living with these ideas and given at least a hundred lectures around the world on the matter, it seemed time to gather together a summary of what I've learned and put it down in book form. My goal here is to present the case for what one might call "social causality" as opposed to "event causality" (although any mention of the term *causality* with its strong whiff of determinism is an automatic red flag). A primary purpose of this book, then, is to make the case for socionomics. Let me say a bit more about what the reader can expect to see in the pages that follow—and what will not be seen.

First of all, the book is a work of advocacy and exposition in support of an hypothesis, what I call the Central Hypothesis of Socionomics. As a work of advocacy the book is definitely "opinionated," as one reviewer of the manuscript not-so-charitably called it. But who would have ever thought otherwise? After all, opinionated is what "advocacy" means, if it means anything. So I make no apologies for that.

The book is also a call to arms for a deeper research program aimed at generating more evidence to validate or reject the Central Hypothesis. Right now, it's simply an hypothesis. The book provides as much ammunition as I can muster to support it. So I make no apologies for not presenting the case for the opposition. One thing this book definitely is not is an academic monograph, where custom demands arguments be presented on both sides of whatever fence the book is straddling. So do not expect to see a case presented here for the conventional wisdom. This book is solely to argue the case for the loyal opposition.

Let me add that I have deliberately written the book in a style very different from the passive, third-person voice so typical of scholarly treatises. This is to make the material as accessible as possible to as wide an audience as possible, as well as to dispel the notion that the only way to expose ideas of substance is to wrap them up in language as incomprehensible as possible.

Another thing this book is not is a treatise on stock market price forecasting. While stock price movements play an important role in the story told here, the story is not about stocks, it's about people. Stock price movements show up only insofar as they are a convenient measure for fluctuations in the mood of a population. So the book is no more about stock prices than a treatise on thermodynamics is a book about thermometers.

A quick scan through the book will show there are many, many graphs. Don't be alarmed. These charts are the same type you see in the financial pages every day in the newspaper. I've only annotated them by inserting different social, political, economic, military or other events at the times when they occurred. So don't be put off by these graphs. They are the "picture" that's better than a thousand words.

OK, now for the good part. It's both a great pleasure and a duty to acknowledge the many friends and colleagues who helped support this work over the past decade with their time and generous comments. For fruitful discussions and suggestions, in no particular order, I thank David Lane, Leena Ilmola, Blake LeBaron, and Jeff Johnson.

The book you hold in your hands has had many incarnations, and numerous friends took time away from pressing daily activities to generously read one or another of those many versions of the manuscript and provide detailed comments and suggestions. For these efforts way beyond the call of friendship, I thank Michael Loescher, Tor Nørretranders, Greg Benford, Ed Binkowski, Gordon Graham, Scott Reamer, Peter Kendall, Rex Cumming, Christoph von Braun, Sally Smith, and Norbert Jesse. I want to express "double thanks" in this regard to my longtime friend Olav Ruud, who kindly volunteered his services as the mythical "target reader" for this book, conscientiously sending me line-by-line comments on the text, as well as suggesting numerous changes in the graphics in order to make the book far easier to understand for a general audience.

Kudos of the first degree are also due the book's editor, Angela Lahee, who took on a project that many editors would have thought was way too much trouble. It's that kind of courage and support that every author of unpopular and/or unconventional ideas wants and needs.

Most of all, I thank Robert Prechter for his wise counsel and continuing support, as well as for the many remarkable graphical depictions of socionomic examples, bringing life to arguments that would otherwise have been made in much drier and far less comprehensible terms.

Finally, readers who would like to send comments, questions, and/or see updates and discussions of the ideas presented here are invited to visit the book's Web site at *www.moodmatters.net*.

Vienna, Austria   John L. Casti
April 2010

Credits and Permissions

*G*rateful acknowledgment is made to the following individuals and publishers for permission to reproduce material used in creating the figures in this book. Every effort has been made to locate the copyright holders of material used here. Omissions brought to our attention will be corrected in future editions.

Elliott Wave International, the Socionomics Institute, and Robert Prechter for copyrighted material noted in the figures themselves as they appear in the book.

Emporis Corporation for Figure 1.1, which appears on the web site www.skyscapers.com.

Robert Prechter for Figures 2.12 and 2.13, which are from *Prechter's Perspective*, P. Kendall, ed. (New Classics Library, Gainesville, GA 2004).

Princeton University Press for Figure 6.1, which is reproduced from Deffeyes, K. *Hubbert's Peak*. (Princeton University Press, Princeton, 2001).

John Wiley & Sons for Figure 6.2, which is reproduced from Sherden, W. *The Fortune Sellers*. (John Wiley & Sons, New York, 1998).

Brad Steiger for Figure 6.4, which is reproduced from Steiger, B., *A Roadmap of Time*. (Prentice-Hall, Englewood Cliffs, NJ 1975).

Overview

The memorable events of history are the visible
effects of the invisible changes in human thought
Gustave Le Bon, The Crowd, 1895

REVERSE PSYCHOLOGY

M ost would agree that John Maynard Keynes was the most influential, if not the greatest, economist of the twentieth century.
In his path-breaking book *The General Theory of Employment,
Interest and Money*, Keynes spoke of the "animal spirits" that impel people
to act outside the scope of deductive, rational thought—action by *feeling
and belief*, not by calculation. The controversial thesis of *Mood Matters* cuts
right to the heart of this most Keynesian of notions, and can be unpacked
into three elementary principles:

I. It is the *social mood* of a group or society, how it feels about the
future, that *biases* the character and the likelihood of the events that
actually occur.

II. The events that do occur have no impact at all on that social mood. In
other words, there is no feedback from events to mood. To put it bluntly

and compactly, *events don't matter* (at least, insofar as the formation of mood is concerned).

III. There are no collective social events occurring *outside* the human social system. So, the notion of an "external" event impacting the social system and thus "causing" people to feel one way and not another is a complete fiction. There is no outside.

Many will accept Principle I; almost no one believes Principle II. So let me start by explaining Principle III and going on from there.

Copernicus' 1543 book *De revolutionibus orbium coelestium (On the Revolutions of the Celestial Spheres)* turned the world upside down (literally!) by arguing the exact opposite of the conventional wisdom of its time, namely, that the Sun revolved around the Earth. *Mood Matters* turns one of the conventional wisdom of today upside down too. Instead of events driving mass psychology, we argue here that the social mood of a population/group is what drives events.

In slightly more technical terms, rather than *event causality*, the book presents the case that the way the world works is through *social causality*. Let me say right now that in the pages that follow I try valiantly to avoid use of the term *causality*, with it myriad philosophical and operational landmines waiting to blow up in the face of the unwary. Instead, I will employ words like *biases, impacts*, or *influences*, which serve equally well for the purposes of this popular exposition without carrying the philosophical baggage attached to the term *causality*. Since the "events-cause-mood" mindset is so pervasive, it's worth considering just how this fiction arose in the first place.

Back in the seventeenth century, Isaac Newton developed his theory of particle motion which even today stands as a model of what a scientific theory of the world should be. According to Newton, a system of particles—planets, billiard balls, atoms—remains in its state of motion unless acted upon by an *outside* force. So we have the idea of a collection of interacting objects (the system) separated from the rest of the world by some kind of boundary (on the nature of which Newton is totally silent). Already in this Newtonian cosmology we see the idea of an external event (the outside force) causing a system to change its behavior. While the Newtonian picture doesn't work very well even in physics, at least not in the realm of very small distances and/or very high velocities, it serves quite admirably to describe what we lowly humans see at the limited scale of time, distance, and speed that characterize our everyday lives.

The huge success of Newton's scheme in describing observations about things as varied as planetary movement and the motion of tides, both phenomena of great practical concern to astrologers, farmers, and sailors, led some to consider publication of Newton's book *Principia Mathematica* as marking the beginning of the Age of Enlightenment. With such pedigree, it's no wonder people began using Newton's theories as the benchmark against which to measure *all* theories of behavior, both in the domains of nature and of humans.

It is but a small step from Newton's views of how particles change their configuration to thinking of humans in a society as being simply "particles" buffeted about by mysterious "outside forces" that give rise to the ever-changing patterns of human behavior. This picture leads immediately to the belief that there are collective social events outside the social group or population, and that those events somehow cause the group to change both how it thinks and how it acts. *Mood Matters* says not so! Just because "everybody" believed the Earth was flat didn't make it so. And just because everybody believes events cause moods doesn't make that so, either.

I see no outside forces; a social group is self-contained, and in fact, there is no room in a social system for something like the "outside." Of course, every model is but a slice of the real world. As a result, the model has to draw a fictitious boundary between the slice of the world it purports to represent and the totality of the real world. The social domain is no different. But where you place the "cut" matters—a lot! And it matters even more in the social realm than in the world of planets, electrons, bridges, and basketballs. It's important to keep this point in mind as we make our way through the book.

This is a good time to return to Keynes and have a quick look at the three pillars upon which his economic theories rest. According to the biographer Robert Skidelsky, the ABCs of Keynes' work are

A. *No External "Shocks"*—the storms battering economic markets today originate in the financial system itself. They are not random shocks hammering the market from the outside, but part and parcel of the normal workings of the economic and financial system itself. We'll say much more about this point later when we come to consider the theories of Hyman Minsky, who argues that the very stability of a market generates the seeds of its later instability.

B. *Networks*—Keynes noted that societies are composed of groups, and, as he put it, are not "branches of the multiplication table." This is another way of stating his "animal spirits" argument that we began this section with. A consequence of his "organicism," to use Keynes' word, is the idea that economic efficiency is not, *in and of itself*, something to be sought at all costs. There's more to life than mere profits.

C. *Long-term Stasis*—a market depressed by an internal collapse can remain in that state for a very long time. So the market cannot be left to correct itself, at least in a time frame that society would find acceptable. Thus, Keynes argued, governments must step-in to right the ship and prevent a recession from developing into a full-blown depression.

We've already stated why Keynes-A is part of our argument. Keynes-C, while being the principle that Keynes is most well known for in the latter-half of the twentieth century, is not something we much ponder in this volume. But Keynes-B certainly is! In fact, it is the single most important aspect of Keynes' work for us, and will be examined top to bottom, left to right, and back to front, by the time we reach "The End." Now back to social groups.

Every group has a social mood, its belief about the future, optimistic or pessimistic, and that mood emerges from forces *inside* the group, not out-side (wherever that may be). That mood in turn strongly biases the collective events that are seen at a later time. In a nutshell, then, this is the thesis we explore throughout this volume.

In *The Wave Principle of Human Social Behavior and the New Science of Socionomics (1999)*, Robert R. Prechter, Jr. proposed what he called the "socionomic hypothesis" of social causality. He summarized the hypothesis as follows: "Social mood is a natural product of human interaction .... Its trends and extent determine the character of social action, including the economic, political and cultural. In other words, the socionomic hypothesis is that mood governs events." In a 2007 article in the *Journal of Behavioral Finance*, Prechter proposed that the "human impulse to herd in contexts of uncertainty" is behind the emergence of social mood.

We can compactly summarize this entire chain of argument in the follow-ing diagram, the Central Hypothesis of Socionomics, which makes explicit the components of this flow of social behavior and action:

THE CENTRAL HYPOTHESIS OF SOCIONOMICS

Herding Instinct $\to$ Social Mood (beliefs/feelings) $\to$ Social Behaviors and Collective Events

So from the proclivity of humans to herd together in groups comes a social mood. That mood then impacts the character and timing of events that we can expect to actually observe. The conventional wisdom argues for reversal of the second arrow in this diagram; socionomists claim the way/why of the world is just the opposite, and that the contrarian view gives a much better account of the actual facts than Newton's view of the world projected onto the social sphere.

There is ample experimental and observational evidence supporting the claim that mammalian brains have hard-wired circuitry impelling the possessors of such brains to tend to cluster together in groups with others of their species. No doubt this herding tendency has great value in the struggle for survival, as individuals can generally defend themselves against predators much better in a group than they can separately. That broad principle is doubly true when it comes to the herding mentality so commonly seen in modern society. In the concrete-and-glass jungles where every day is a battle against merciless "predators" of all sizes, shapes, colors, and forms, there is enormous "denial-of-responsibility" value in following the herd. And if there is one survival principle that stands above all others in today's world, it's being able to CYA (aka avoid responsibility).

It is by no means well established that the herding instinct *on its own* can account for the creation or destruction of an overall social mood. After all, this impulse is the property of an individual brain, and whether that is the key property leading to the formation of a crowd dynamic such as a mood in a population is an open issue. It seems the way to bet. But much more empirical work needs to be done under controlled circumstances before we can comfortably accept this as a fact. For now, it's simply a good working hypothesis.

Strictly speaking, of course, there are several "micro moods," each representing the beliefs about the future held by some subgroup of a population. But as already noted, scientific models necessitate simplifications from the actual situation in order to make progress in understanding a broader picture. So for the purposes of this book I will assume that whatever group we're speaking about—the population of a city, country, or even the entire world—has just a single mood at any given time for a given time scale.

We should also note that the arrows in the diagram suggest there is a flow of information from left to right as we make our way across the diagram. And that is indeed the case. But what the diagram does not show is the *time* required for that information to flow. So lest the reader infer that the flow is more or less immediate, I want to point out that this is most definitely *not* the case. And, in fact, the time lags involved in moving information from one part of the diagram to the next is an integral part of our overall story.

Let us make one last bow to Keynes by citing the following statement made in the preface to *The General Theory of Employment, Interest and Money*: "A monetary economy is one in which *changing views about the future* [emphasis added] are capable of influencing the quantity of employment and not merely its direction." Robert Skidelsky calls this statement the "core of the Keynesian revolution." It serves equally well as the core of the ideas presented in this book, too.

In the world of finance and economics, it's now well established that the financial markets are a leading indicator of macroeconomic events, such as unemployment rates, money supply, and the like. We will see later that the financial markets provide an excellent "meter stick" by which to measure the social mood. The social mood then becomes a leading indicator of macroeconomic events as well, just as the Central Hypothesis would suggest. The more wide-sweeping argument of this book is that the social mood is not just a leading indicator of macroeconomic phenomena, it is a leading indicator of *all* collective social phenomena.

I don't want to start an extended argument in this brief overview in support of the Central Hypothesis. After all, that's the point of the entire book. Instead, let me now briefly outline the chapters of the book, returning in the pages that follow to the counterintuitive claims I've made above in Principles I–III.

CHAPTER OVERVIEW

The basic plan of the book is to begin in Chapter 1 with several examples illustrating our thinking about the relationship between social mood and events. Chapter 2 then proceeds to examine in detail the main content of the Central Hypothesis shown in the diagram presented earlier. The next three chapters provide an extensive array of empirical examples showing the Hypothesis in action for events unfolding on short, intermediate, and long timescales. The final chapter brings everything together, and offers the reader a set of tools for getting a glimpse at how the world is likely to look in various domains of life during the months and years to come.

The book also contains two appendices. The first is a short, self-contained explanation of the Elliott wave model for projecting financial time-series data into the future, a method I find especially useful for reasons that will become apparent as we procede through the book. The second is an outline for the type of research program I believe is needed to turn the Central Hypothesis into something approaching a bona fide scientific theory for the "why" of social change.

Contents

Contents

Chapter 1

THE *UN*CONVENTIONAL WISDOM

REACHING FOR THE SKY

A couple of years ago when the Great Asset Mania was raging in full force, I saw an article extolling India's new-found position in the world scheme of things, one of the many outgrowths of the globalization of the world's economy, which as it happened was just then beginning to roll over. Somewhere in the middle of this piece was a statement by Delhi architect Hafeez Contractor (what a great name for an architect!), who was obviously caught up in the euphoria of India's move onto the world's center stage. He was speaking out for building the world's tallest building in Delhi, stating "It is about status. It is about glorification. It is high time that people started realizing that we too are a great nation." While I think most would agree that India *is* a great nation, even without Mr. Contractor's cheerleading, his statement underscores a point worth pondering.

Figure 1.1 shows the world's five tallest buildings as of late 2009. Three of the five buildings are in small Asian countries or the Middle East. Moreover, all have been built within the past decade. And this is not just for the very tallest buildings, either. It turns out that 19 of the 25 tallest buildings in the world share this geographic and temporal property. So what's going on here? Why are these "little guys" on the world's geopolitical/economic stage all of a sudden trying to become "big guys"?

We need look only slightly below the surface of Mr. Contractor's *cri de coeur* to find the force driving this flurry of skyscraper construction in the newly-minted economic powerhouses of Asia and the Middle East. It's the very same principle underlying what at the individual level is often informally described as the "Short Person's" personality.

J.L. Casti, *Mood Matters*, DOI 10.1007/978-3-642-04835-7_1,
© Springer Science+Business Media, LLC 2010

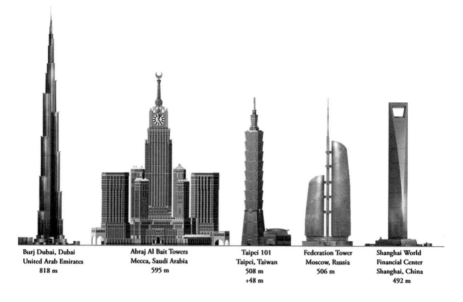

| Burj Dubai, Dubai United Arab Emirates 818 m | Abraj Al Bait Towers Mecca, Saudi Arabia 595 m | Taipei 101 Taipei, Taiwan 508 m +48 m | Federation Tower Moscow, Russia 506 m | Shanghai World Financial Center Shanghai, China 492 m |

Figure 1.1   The world's five tallest buildings as of late 2009.

Psychologists have observed for a long time that human beings, particularly men who are noticably shorter than average develop personality traits like assertiveness, loud voices, and bustling, take-charge behavior in order to ensure that others notice them. We see the same phenomenon at work at the level of countries. And what better—and cheaper—way to proclaim to the world "We're here!" than to build the world's tallest building? Basically, such a tower is a monument to a society's belief that their future will be just like today—only brighter. So instead of building something that fits the social needs of today, they build something that attests to their vision of tomorrow, tomorrow that rises onward and upward, presumably forever.

But trend following can be a tricky business, as the chart in Figure 1.2 dramatically illustrates. Here we see the Petronas Towers, Taipei 101, and Burj Dubai, each of which was the world's tallest building at the time of its completion. As a measure of the changing level of optimism for the future in each country, the figure also shows the local stock exchange index. In each chart, arrows indicate when construction of the local monolith began and when it was completed (except Burj Dubai, which is scheduled to be "topped out" in early 2010, and now renamed Burj Khalifa to honor the leader of Abu Dhabi, who bailed out the fast-sinking Dubai economy in December 2009).

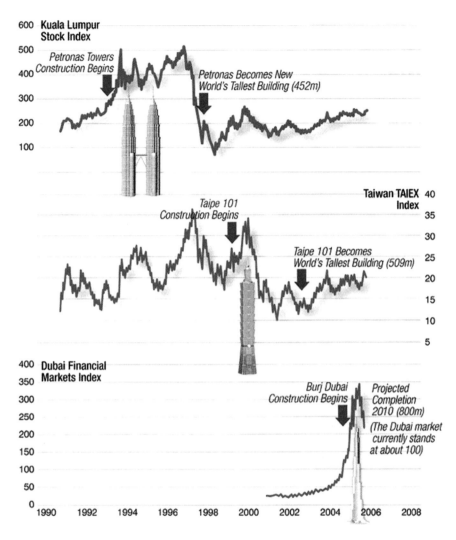

Figure 1.2 Skyscrapers with local stock indexes (Adapted with permission from the *Elliott Wave Financial Forecast, April 2006.*).

History has shown an eerie correlation between tallest-building initiatives and financial debacles. It seems the urge for architectural monuments to human hubris is about as reliable an indicator as you can find for financial trouble. Economist Mark Thornton states that skyscrapers tell us a lot about a country's technological capabilities, ambition, economic wealth and need to project itself on to the world stage. He says, "[It's] these features that make skyscrapers, especially the construction of the world's tallest building, a salient marker of 20th-century business cycles."

The right thing for the reader to wonder at this point is why we're using the stock exchange index as a way of measuring the overall feelings of a population about the future. We'll talk much, much more about this in the next chapter. For now, it suffices to note only that the stock exchange reflects bets people make about the future on all timescales. So the index certainly says *something* about how a society feels about what's likely to turn up next. There are many big issues surrounding Robert Prechter's suggestion of using of the stock index as a social mood meter, or "sociometer," including the fact that even in a stock-crazed society like the United States not everyone is directly engaged in buying and selling stocks. But that's for later. For now, let's get back to our skyscraper story.

As noted by Peter Kendall in 2006, in each of the three cases shown in Figure 1.2 a funny thing happened on the way to the top. At the time construction began on each of the buildings shown, people's optimism for the future knew no bounds and the social mood meter, the local stock index, was soaring off into outer space. But skyscrapers aren't built overnight. And without exception by the time the building was completed the index was lower than snake hips, indicating that the local citizenry was deeply pessimistic about the future. While I don't have space here to show this story for earlier efforts like New York City's Empire State Building in the 1930s, I hope the reader will take my word for it that the very same phenomenon was at work there too.

The basic message is that when a society is optimistic about the future and is trying to get recognition on the world stage, an easy way to broadcast their "We've-arrived" message is to build the world's tallest building. But by the time that building is complete, the population has undergone a sea change in its beliefs about the future and the national mood has sunk back to a level far below where it was when construction began.

What we see at work here is what might be called The Skyscraper Index. One implication of this Index is very clear and simple: If you see a country starting to build what it claims will be the world's tallest building, the time has come to start bailing out of stocks in that country's market as fast as you can! In short, when everyone starts believing the future is rosy, it's already very late in the day for the underlying positive social mood. By then that mood is mature and a rollover to the negative is imminent. We'll see a lot more of this kind of contrarian thinking in the chapters to follow.

4

For far-sighted (and thinking) readers, it's worth noting that South Korea has just announced construction of the world's *second* tallest building, Lotte Supertower 123, to be built in Seoul. This will be the tallest building in Asia, and only exceeded in height by Burj Khalifa. Excavation of the site was completed in 2009, and the tower is scheduled to be topped out in 2014. For visionary investors it wouldn't hurt to start looking for the top of the Korean stock market . . . soon.

Before moving on to other matters, this is a good moment to emphasize that the goal of this introductory chapter is to present ideas relating social mood to *collective*, not individual, human events, and to suggest a direction from mood to events that runs directly counter to the conventional wisdom claiming that things unfold exactly the other way around. A natural consequence of this topsy-turvy argument is that the reader schooled in conventional wisdoms as to how the world works is going to want to engage in debate at almost every stage of our presentation. That's all to the good, since that's what this chapter is about: raising questions—but not necessarily answering them. The remainder of the book is devoted to providing as full, complete, and honest answers as we can give at this stage of our knowledge. As we proceed, I'll try to point out what I believe are the most important questions that we'll be looking at later in the book. So by the end of our story there will definitely be answers. But I don't claim there will be a one-to-one correspondence between the two! Just something better, a lot better actually, than what you'll have in mind at the end of this preliminary chapter.

Now let's have a look at something a bit more serious than skyscrapers but still conveying pretty much the same message. And that message is that the kinds of things that happen when a group is optimistic about the future are, generally speaking, qualitatively *very* different from the sorts of events that tend to occur when the dominant mood is pessimistic.

THE ART OF THE POSSIBLE

Conventional wisdom has it that political trends are a key determinant of the stock market's gyrations. As an election approaches, commentators endlessly debate the effect the outcome of the election will have on stock prices. Investors weigh up which candidates will influence the market to move up or down. Statements like "If Jones is elected, it will be good for the market, but Smith's election will cause stocks to tank" are common.

If this causal relationship were even approximately correct, there would be evidence that a transfer of power from one party's leader to another affects the social mood, hence, the stock market, in some very specific ways. There would also be evidence that certain political parties or policies reliably produce bull or bear markets. There is no study showing any such connections or correlations. On the other hand, its not hard to see just the opposite at work.

A strong and persistent trend in the stock market dramatically biases whether an incumbent president or the incumbent's party will be re-elected in a landslide or defeated in one. In all cases where an incumbent remained in office in a landslide, the stock market's trend was up at the time of the election. In all cases where an incumbent lost in a landslide, the stock market's trend was down—as it was prior to the 2008 election that swept Mr. Obama into the White House.

Again using the stock market index as a surrogate for how the American citizenry rates the future, we find that there is not a single case in which an incumbent was re-elected despite a deeply falling stock market or was defeated in a landslide despite a strongly rising market. According to Robert Prechter's analysis, the conclusion is clear:

> When a population is optimistic about the future so that the social mood waxes positive as reflected by persistently rising stock prices, voters wish to retain the leader who symbolizes their upbeat feelings and who they presume helped cause the conditions that gave rise to their feelings. On the other hand, when the social mood turns negative as reflected by persistently falling stock prices, voters throw out the incumbent who symbolizes their downbeat feelings and who they presume helped bring about the conditions leading to their pessimistic outlook. The actual political policies of the incumbent and the challenger are irrelevant to this dynamic. The key is a desire for change per se, not for any particular type of change. The conventional wisdom that the president and/or his party drives the stock market, has no explanation for reconciling the relationship between these phenomena.

To illustrate, in November 2004 the overall mood of the American populace as measured by the Dow Jones Industrial Average (DJIA) was rather upbeat. This mood then translated into a not-entirely-landslide victory for George W. Bush. But it was a victory, nonetheless. This line of argument

6

certainly gives cause to ponder the outcome of the 2012 US presidential election. If the social mood declines between now and then, the Democratic incumbent Mr. Obama is very likely to become a one-term president and go the way of Nixon, Hoover, Carter, and all the other political "dodo birds" of the species *politicus americanus*. But if the markets miraculously recover and are surging upward in November 2012, then the return of a Republican president is very unlikely. That's a big "if", of course, as its still some way to the autumn of 2012.

A small straw is already blowing in the wind, though, as a poll taken at the end of August 2009 showed that Americans are losing their faith in President Obama, especially in his seeming inability to curb the growing federal budget deficit and to pump up the moribund job market. The fraction of the population that said they believed Mr. Obama would make the right decisions for the country had declined to 49% from the 60% at the 100-day mark of the Obama presidency. A few months later, the situation was no better as Obama's popularity slid in January to 47%. And the Congress was doing no better. Over 60% of the respondents said they were not satisfied with one or more of their representatives, and 29% said, "Representatives? We have representatives?" But, again, it's a long time to the end of 2012, and polls like these are well known to move all over the place as events come and go.

To conclude this example, our claim is that an increasingly negative social mood impels voters to magnify politicians weaknesses, minimize their strengths, perceive that old political styles have failed, and eventually kick out the incumbent at the next election or even before it occurs. The situation with Governor Gray Davis in California a few years back is a good example of this shift in mood threatening an incumbent politician's job. A bear market raged for 3 years from 2000 to 2002. So voter hostility had to erupt somewhere. And it did, leading to Davis' recall and the election of "The Terminator" as the governor of California, as we will discuss in greater detail in a later chapter.

Summarizing the overall lesson, the conventional analyst asks, "How do political leaders' actions cause a change in their popularity (social mood)?" The social mood analyst asks just the opposite question: How does the social mood affect voters' perceptions of political leaders? *This* is the relevant question, and it is supported by strong statistical studies cited in the References for this section.

To conclude this introduction to social mood and its impact on events, here is a global-level illustration.

CHAPTER 1

FROM GLOBALIZATION TO LOCALIZATION

Unlike skyscrapers which are an inherently local phenomenon, confined to a particular geographical space, the not-so-currently-trendy idea of globalization, which views the world as one gigantic marketplace for fulfilling the dreams of humankind, unfettered by the inconveniences of restrictions on the flow of capital, labor, materials, or ideas, is another collective social phenomenon that is on the road to coming undone.

In the fall of 2006, I was invited to make an address at the annual meeting of the European Futurists Society in Lucerne, Switzerland. To liven things up a bit, I chose as the title of my presentation "The Decline and Fall of Globalization," a theme not at all in harmony with the rosy optimism on offer from the overwhelming majority of presentations at that event. In point of fact, I would say my talk probably qualified as the meeting's least popular presentation as few, if any, in the audience had anything supportive or constructive to say about it.

As fate, or perhaps divine justice, would have it, though, just two short years later the organizer of this annual event, Georges Roos, phoned me a couple of weeks before the event and asked if I could come and say a few words at the 2008 meeting about how I came to ideas so strongly rejected two year's earlier. I'll have more to say about this matter later in the book. At this juncture I only want to record the fact of this "tale of two presentations," in order to enshrine in print my sense of satisfaction in being able to return to the very same podium and say (discreetly, of course), "Do you remember what I told you two years ago?"!

In its modern incarnation globalization had its birth at the 1971 World Economic Forum meeting in Davos, Switzerland. That was a time when the social mood was escalating upward at a dramatic rate. A few years later, the G8 was formed, followed by the North American Free Trade Agreement (NAFTA), and later the formation of the World Trade Organization (WTO). Since the driving force behind globalization is to a large degree the American corporate community, we look to the DJIA from 1970 to 2004 as an indicator of the overall global mood for this example, since the New York Stock Exchange is still about the closest thing we have to a global financial market. Every single milestone in the path to globalization from the launching of the basic idea at Davos to the formation of the World Trade Organization in 1996 to China's joining of the WTO in 2000 took place at a peak in social mood, as illustrated in Figure 1.3. Here I've inserted

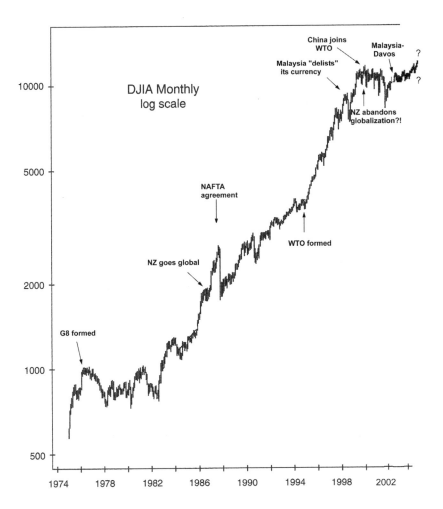

Figure 1.3 The road to globalization, 1970–2004.

question marks as to how things might unfold after 2004. I guess by now (spring 2010) we all know the answer to *that* question!

From 1975 to 2000, the global social mood was rosy enough to emit heat. In such times, the types of events we expect to see are those that can be labeled with everyday words like *unifying, joining, welcoming*, and *expanding*. Sad to say, the picture shows that by 2000 this global mood was already starting to flatten out. While the process actually took about five years to really get going, we can with some confidence say that we're now in the early stages of a decades-long decline, one that's likely to lead to social behavior and events of just the opposite character: Localization instead of globalization, fragmentation instead of unification, xenophobia instead of welcoming.

All of these types of behaviors are already evident in the pages of your favorite daily newspaper (if it still exists!) or on the Internet. Here are a few headlines to ponder from a single six-month period in 2007:

Unexpected Results: Globalization Has Widened Income Disparity. *Wall Street Journal, May 24, 2007.*

Globalisation Backlash in Rich Nations. *Financial Times, July 22, 2007.*

You Can't Stop Us Coming. *Daily Mail, London, January 2, 2007.*

The Rich Get Richer as Workers Struggle On. *New Zealand Herald, January 26, 2007.*

G4 Talks Collapse, Throw Trade Round into Doubt. *Reuters, June 2, 2007.*

Trade Talks Fail to Get Doha Plan. *BBC News, May 18, 2007.*

Do any of these headlines sound like they describe a story you'd label as being about "joining," "embracing," "unifying," or "tolerating"? And they're not about minor technicalities or marginal aspects of globalization either, but about the very foundations of the whole process itself: income balance, free movement of labor, reduction/removal of trade tariffs, and the like. The remarkable thing about headlines like these is that there is nothing remarkable about them! They are the type of headlines we now see almost every day. Even more ominously, they came at a time when the global social mood as measured by the DJIA was showing major stresses and strains, retreating big time from its all-time top on October 11, 2007.

The reason for the claim we made above is that this process is just starting a decade(s)-long decline. As it took the better part of three decades for globalization to really "take off," it's definitely a long-timescale phenomenon. While the unwinding of a trend generally runs faster than its development (since it's fear that drives the unwinding, not greed, and fear is by far the stronger emotion), we won't see the old global spirit really reassert itself for a period that's at least commensurate with what it took to get to the peak in 2000. Since the real downward pressure didn't really get started until 2007, it's not much of a call to argue that it won't bottom out in much less than 10–15 years—and maybe more. We'll have more examples and arguments supporting this claim in the appropriate chapter later in the book.

I think it's self-evident that what we call "globalization," the disappearance of barriers to the free flow of everything involving trade—people, money, information, ideas, and materials—is a collective human social activity. So under the assumption that such activities are biased by the overall social mood of the world population, we're forced to conclude that the phenomenon of globalization itself is also biased by that same mood. Futher details on the trend toward localization can be found in the References for this section.

A REVOLUTION IN THE MAKING?

Our examples—skyscrapers, elections, globalization—while very different in character and geographic scope, share several features that serve to focus much of the story we tell in this book. These commonalities include

- *A Group of Individuals*—Two of the examples involved the individuals constituting the population of a country, while the third was the entire global community. In other cases, the group might consist of the residents of a particular city, members of a common interest group like an Internet Forum, or even just the employees of a firm. But in all cases these individuals each have some degree of belief/feeling about the future.

In addition, the group is heterogeneous in the sense that some members of the group are better able to "sell" their views to other members, while some members are more susceptible to "buying" than others.

- *Herding Behavior*—Whatever the group, its members have a tendency to herd together into collectives, thus forming a *social* group rather than being simply a collection of isolated individuals. This "herding" instinct was almost surely hardwired into mammalian brains many eons ago as an evolutionary survival trait. And so it remains to this day. Each of our examples involves a collective taking some kind of action leading to a specific event, be it the construction of a building, the results of an election, or a shift in the way of doing business.

- *Social Mood*—For compactness, let us adopt Prechter's term and call the way a group feels about the future its *social mood*. In some poorly understood fashion, the social mood emerges from the moods of the individuals constituting the group, together with the manner in which they interact. Thus, the social mood is an emergent property of the group *as a whole*. But its measure is nothing so elementary as simply adding up the feelings of each member of the group. The interactions matter—a lot. So does the fact that some people are just more adept in shaping the overall social mood than others.

As noted in the Overview, the social mood of a group is not a homogeneous property; there are actually several moods, one for each identifiable subgroup. Often these "micromoods" coincide. But sometimes not. For the sake of both simplification and exposition, as well as the fact that these micromoods often "cancel out," we'll adopt the position in this book that there is a single, dominant mood for a group at any moment in time, and employ the term *social mood* to characterize that particular belief of the group about its future.

- *Events and Actions*—The social mood biases the types of events that are more or less likely to take place within a particular time frame. So while a great variety of events are possible in the sense that they are not ruled out by the laws of physics or humans, not all such possibilities are equally likely. Some are simply more likely than others. And what is or isn't likely at a given time is strongly conditioned by the view the population holds at that time about the future.

For example, when a population is wildly optimistic about the future, it's far more likely to take actions reflecting that optimism, such as building the world's tallest building. By way of contrast, when a society fears the future, it's much more prone to look for things like a change of leadership in the political arena.

A note on terminology: The term *event* is used here as if everyone knows and understands what it means. The matter is not so clear-cut as one might think—or wish. Some events have a definite moment of occurrence, such as the outcome of an election or the outbreak of a war. But how does one date the "occurrence" of globalization? Or rising skirt lengths? Or the success of a vampire film? Such events are sort of smeared out in time, and the actual moment of their occurrence is often unclear. I raise the point here just so the reader knows that I'm aware of the fact that an "event" is a rather ill-defined notion, somewhat like the definition of pornography given by a US Supreme Court justice some years back: I can't define it, but I know it when I see it. So we will use the term in this "pornographic" sense throughout the book, trusting to the reader's intuition and judgment as to what it actually means in any given situation.

We can compactly summarize the above paragraphs in what we will term the Central Hypothesis of Socionomics expressed in the diagram below, which we saw earlier in the Overview section. But for something as important to the focus of our story as this, repetition is no sin. It shows the components of the socionomics argument that we'll explore in the pages to follow.

THE CENTRAL HYPOTHESIS OF SOCIONOMICS

Herding Instinct → Social Mood (beliefs/feelings) → Social Behavior and Collective Events

This flow from individual herding instincts through to collective social actions and events via the medium of social mood has been the focus of a series of books and articles by financial guru and social theorist Robert Prechter. By design, the terminology *socionomics* introduced by Prechter differs from the word *socioeconomics*, an academic field that mixes sociology and economics. Although the two areas of study both center about issues involving social behavior and economics, digging a bit deeper into both one finds that the two lines of thought have at most a very tenuous relationship. So whenever the term *socionomics* apppears, the reader should remain alert to this distinction between the two fields of study as we wend our way through the story unfolding at that particular moment. In this volume it means one thing and one thing only: the flow of influence encapsulated in the diagram presented above.

An important caveat: The arrow from Herding Instinct to Social Mood and the one from Social Mood to Behaviors/Events does not mean that the first *causes* the second. The reader should interpret these arrows as shorthand for words like *predisposes, impacts,* or *biases,* each of which conveys much better the meaning I have in mind. Thus, the prevailing social mood creates a bias for social events that are more or less likely to occur. As Prechter puts it, "Social mood is responsible for the character of social action." But it doesn't necessarily "cause" those actions. The next chapter will make this distinction a lot clearer (I hope!). Now let's take a quick look at the main issues the book will address.

THE BIG QUESTIONS

The enigmatic Austrian philosopher Ludwig Wittgenstein once remarked, "The solution of the problem of life is seen in the vanishing of the problem." The vanishing of the problem of why we see what we do in the social domain and not see something else lies in a much deeper understanding of each component of the diagram shown above. This will be our leitmotif throughout the remainder of this book.

To use Wittgenstein's suggestive terminology, the principal questions we want to have "disappeared" can be grouped into four broad categories: timescales and time lags, formation and measurement of social mood, feedback from events to mood, and forecasting and anticipation of behaviors and events that are more or less likely, given a prevailing social mood. We briefly outline the overall thrust and scope of each of these categories by way of preparation for the full-scale assault on these matters to be taken up in subsequent chapters.

TIMESCALES AND TIME LAGS

It's pretty clear, I think, that social events of different types require different periods of time to unfold. For instance, the time between breaking ground on a skyscraper and its actual completion is a few years. By way of contrast, the full unfolding of the globalization mania took a couple of decades or thereabouts to run its course.

So it is when we're thinking about social mood, too. If I asked how you feel about the future, the right reply would be, Which future? Do you mean How do I feel about tomorrow? Next month? Next year? It's easy to imagine your answer might be very different for each of these time periods. Just as with events, then, the way a society feels about the future also has its own set of timescales.

Given the fairly obvious fact that when looking at the social mood and its impact on what events are more or less likely to occur, we have to take into account the unfolding time of the event and match that timescale with the timescale of the social mood serving to bias the occurrence or non-occurrence of that event. Again using the skyscraper example, it wouldn't tell us anything useful about the possibility of a skyscraper going up in Dubai if we looked at the Dubai stock index price movements on a daily basis. This timescale is simply too short to see the social forces at work leading to the construction of a skyscraper. More appropriate would be to look at the index on a monthly, or even quarterly, basis if we want some insight into the possibility of a world-class skyscraper to come shooting up out of the desert.

To oversimplify the situation a bit, we can identify three primary timescales for the social behaviors of interest in this book. They are

- **Short-term Events**—*A few weeks/months to a few years*: Here we find mostly events in the realm of popular culture. The types of films and books that are popular, the types of fashions that come into vogue, and the sort of music that people like are all examples of events unfolding on this short-term timescale. For these cases, one would want to read the "social mood meter" on a more-or-less weekly basis to get some idea of what's likely to be "the next big thing."

- **Medium-term Events**—*A few years to a decade or two*: In this category are things like shifts in political ideology, changes in macroeconomic trends, the emergence of processes like globalization, or the outbreak of war. Here the mood meter should probably be read quarterly to annually to get a picture of what's unfolding and when.

- **Long-term Events**—*A few decades to a few centuries*: Major geopolitical shifts and the rise and fall of great powers are events that are best seen in this time frame. The appropriate time step for reading the mood meter for tectonic shifts in the social order of this magnitude ranges from a few years to a decade or two for events unfolding over a century or more.

Probably the biggest question we have to address in the category of timescales is to identify what time frame is really the most appropriate one for a specific event. This is the Time-Lag Problem, which centers upon the time it takes from a shift in social mood to the occurrence of the events and actions reflecting that change of mood.

Earlier, we mentioned the difficulty in pinning down just when an event can be said to have taken place. The timescale taxonomy just presented is very crude and gives only a rule-of-thumb indication of the relevant time period for an event. There are many borderline cases, and it's most likely that we'll have to experiment a bit with different time frames in order to get one that seems to work for a specific event of interest.

THE FORMATION OF SOCIAL MOOD

A good way of thinking about the formation of social mood in a society is to envision how a flu epidemic develops. You have a collection of people, some of whom are infected with the virus to one degree or another. There is also a part of the population that is not infected—but who are susceptible to catching the flu from someone who has it and then transmitting it to others before they die or recover. (For simplicity, we neglect those who are immune as they are irrelevant to this example.) All these people are interacting in some way, each "trying" to infect others and/or avoid being infected himself or herself. If a critical mass of people get the flu, and if the interaction pattern is sufficiently densely-connected, then a wave of infection spreads through the population until it finally runs its course. In short, there's an epidemic.

Presumably, this same sort of mechanism is at work in the formation of the social mood in a population. The only real difference is that now it's ideas and beliefs about the future that are passed around from one brain to another instead of bacteria and viruses moving from one body to another. Moreover, in the social environment the beliefs of some people/institutions are just much more effective in "coopting" your brain than others. And some people's brains are more easily influenced than others. So just as with a biological infection, the way people are connected to each other and how these connections facilitate the exchange of views about the future deeply influence whether there will be a jump from one mood polarity—positive or negative—to the other.

This description is sometimes called the "contagion" theory for herding and group opinion formation. Some, though, argue for a more "interactionist" theory, saying that your psychological state is not like a virus or bacteria that you can "infect" someone with. Rather, we get all sorts of cues from others as to how they feel by listening to what they say, observing their body language, and the like. In other words, by interacting and observing them. These interactions are then the raw material by which people form their views. Whether there's a difference that matters between these two models of mood formation is yet another major research area calling out for attention.

A huge question, of course, is how *exactly* this mood formation takes place in a highly heterogenous population. That is still a deep, unanswered research question. It's reasonable to assume that the answer is bound-up with the types of "transmitters" and "receivers" in the society and the way they interact. But that's not saying much. What matters is the specific *way* these factors line up to create a sharp change of mood, a kind of "phase transition" from one social mood to another. And note here we're talking about a *change* of mood, not the creation of the mood itself. The social mood always exists; it doesn't have to be brought into existence. So our focus is on how changes take place in the mood polarity, not on how the mood itself comes about. The mood is always there.

THE MEASUREMENT OF SOCIAL MOOD

How can we measure the mood? It's difficult to think of how an actual science might emerge from the Central Hypothesis of Socionomics without being able to actually quantify the mood of a population at any point in time, not just observe or sense it somehow. In the examples given earlier, we've taken the financial market averages as our "sociometer" for measuring how people in a society feel about the future. In the next chapter we'll see that there are a number of very good reasons for doing so. We'll also look at alternatives. It is another open question as to whether there might be ways to measure the mood that are more direct and more effective than price movements in financial markets. Public-opinion surveys and social networks like Facebook and Twitter immediately come to mind.

FEEDBACK FROM EVENTS TO SOCIAL MOOD

Common belief has it that when an event takes place, especially a dramatic event like a presidential assasination, a terrorist attack, or the collapse of a financial institution, the occurrence of that event changes people's feelings about the future. In other words, it affects their mood and thus the mood of the entire society. Or so goes the conventional wisdom, anyway. This sense that events impact feelings seems so strong, in fact, that most people don't even question it but regard it as a taken-for-granted background reality to the way things are, always have been, and always will be. The only problem with this "everybody-knows-it" belief is that there is not a single piece of actual evidence to support it. Just like a few centuries ago when everybody "knew" the Earth was flat, the alternative that perhaps it really isn't flat was simply dismissed out of hand as the ravings of a crank or a madman.

Once we've figured out how to measure social mood, we'll be able to properly test this notion of feedback from events to mood. Since the Central Hypothesis contains no arrow pointing back in this direction, even the least attentive reader will be able to guess the position we take in this book. What's interesting, though, is not our position but the *reasons* and the *evidence* for that position. For that, stay tuned.

ANTICIPATION AND FORECASTING OF EVENTS

The distinction between anticipating an event and forecasting it is a subtle, but important, one. To illustrate the difference, consider an event like the recent collapse of the US housing market. On the basis of signals from the mortgage and housing industry in 2005 that strongly pointed to a wildly overheated market, one might have reasonably *anticipated* a major turndown— soon. So among the many possible paths the housing market might have taken, a major fall in prices seemed the most likely. But this is not the same as an actual *forecast*. For that one would have to make a statement like "With 90% likelihood the housing market will go into freefall beginning in Fall 2006." This involves a declarative statement that something *will* occur with a given likelihood and that the occurrence will take place at a specific time in the future.

Our goal in this volume is to use the concept of social mood as a kind of telescope with which to look into the future with just a bit more precision than is generally the case. To be blunt about it, I don't believe there is a person/method living, dead, or yet to be born that can consistently and reliably forecast *specific* social events. But I do believe that one can anticipate the "climate" (= social mood) within which events will unfold, and that this climate biases to a large degree the character of the type of events that are more or less likely to take place. As the French molecular biologist Jacques Monod put it in the title of his popular work on evolutionary processes, social events are the product of chance and necessity. What we're after in this book is the "necessity" part. Here's another everyday example to fix the idea.

Think about a game of American football, where we have a collection of 22 players, each taking actions in accordance with specific ways (that is, rules) for playing his or her position. These actions are themselves circumscribed by the overall rules of the game. Moreover, the interactions of the players takes place on a playing field, which itself biases the likelihood of certain types of events. For instance, if it's a rainy day and the field is muddy, fumbles and dropped passes are far more likely than if the field were dry. So events emerging out of the interactions among the players, such as touchdowns, field goals, missed tackles, lost footing, and false starts are all biased by the condition of the field.

But, of course, this is just an analogy. And like all analogies it is flawed in the sense that the actions of the players themselves do not give rise to conditions like wind, rain, sunshine, and the like. The analogy would be tighter if that were not the case, since the social mood, which is analogous to the playing field, is actually created by the population. A weak step in this direction would be to note that the actions of the football players do sometimes impact the condition of the playing field, since if the field is wet and soggy to begin with the players running about on it tear up the turf and thus predispose certain types of events, like the loss of footing that would have been less likely had the field been dry and the footing firm.

In the world of human activities, the social mood, the *Zeitgeist*, serves as the playing field. And on this field, individuals, institutions, and governments interact in ways that cannot really be understood in any precise way. The actions and interactions of these "players" are the chance part of the situation; the social climate is the necessity. And it's this climate that pushes events in one direction and not in another. Our goal in this book is to understand just how that *Zeitgeist* unfolds on many different timescales, and how that unfolding gives rise to why we see the events we do and not see something else.

Our first way station on the road to the future is to look much more deeply into social mood and its influence on behaviors and events. In the next chapter we examine where the social mood comes from, how it changes, and the various ways we might measure these changes.

Chapter 2

WHY "STUFF" HAPPENS

THE START OF SOMETHING BIG

*I*n the autumn of 2001, the financial collapse of the Enron Corporation hit the front pages of virtually every newspaper in the world. At the time it was the largest bankruptcy in US corporate history. But as scandals go, it turned out to be just one of many accounting "irregularities" that numerous American corporations had been practicing for the decade or more during the runaway bull market that began in the early 1980s.

By far the most interesting aspect of the Enron collapse, though, was the public's reaction to the event. Basically, both the financial and general press promoted the view that the Enron accounting revelations had deeply discouraged investors, thereby *causing* a crisis of confidence on Wall Street. In essence, the conventional wisdom of the chattering classes was that Enron's collapse *generated* a negative social mood, which in turn led to a lack of investor confidence in the market.

Notice the italicized words in the preceding paragraph, "causing" and "generated." They suggest a direction of causality in social events, one that is so deeply hardwired into the collective subconscious that to question it is akin to challenging our taken-for-granted reality as to the way the world works. It's useful to consider the Enron situation as an entry point into the fundamental questions of what is the social mood and how it influences social events and actions. Robert Prechter discussed this Enron event at length in material cited in the chapter References. Let's review his argument.

A headline in *USA Today* in the spring of 2002 captured succinctly the mainline view of Enron social dynamics when it proclaimed to the world, "Scandals Shred Investors' Faith." The implication of this headline is that

J.L. Casti, *Mood Matters*, DOI 10.1007/978-3-642-04835-7_2,
© Springer Science+Business Media, LLC 2010

the market was moving along just fine—until the Enron revelations shattered people's confidence in stocks. If this line of reasoning were even approximately correct, one would have expected the market to experience a precipitous selloff *following* the revelation of Enron's accounting shenanigans and the company's consequent bankruptcy filing. If you're tempted to believe this fiction, have a look at Figure 2.1, showing the daily price movement of the Standard & Poor's 500 Index (a benchmark measure for overall market performance) from the year 2000 onward.

The figure shows that in the 18 months *preceding* the Enron scandal, the market declined 39%. *After* the scandal broke in 2001, the market actually rose more than 10%—and stayed up at those levels for nearly a year afterwards. So the actual facts of the matter fly completely in the face of the notion that the Enron collapse "spooked" the market; in point of fact, the flow of events ran in just the opposite direction. Here's what actually happened.

The declining market from January 2000 onward put enormous pressure on Enron's ability to use its high-flying stock price as leverage to secure

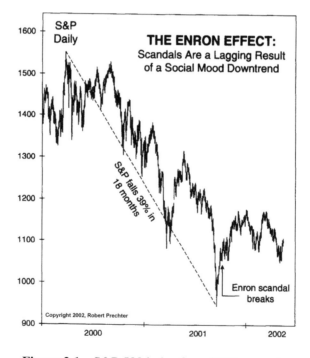

Figure 2.1 S&P 500 index from 2000 onwards.

loans to support the firm's accounting legerdemains. The decline of the stock price and the consequent drying up of the firm's lines of credit then led to the collapse of the company—and the subsequent scandal—when regulators and creditors began digging into the company's books. The increasingly negative social mood also whetted the public's appetite for scandal, recrimination, and punishment. A scapegoat was needed and Enron was the perfect candidate.

So contrary to the wisdom of the time, a more viable line of reasoning is that it was the negative social mood, as reflected in the dramatically falling stock market, that led to the Enron collapse, not the other way around. In this view, investors were not depressed at all as a result of the Enron collapse; rather, they were *already* depressed for the preceding 18 months as Figure 2.1 so graphically illustrates. In a very definite sense then, it was this negative tone in investor psychology that led to Enron's collapse. And if it had not been Enron it would have been some other firm employing similar magical accounting procedures—which indeed turned out to be the case, as witnessed by the subsequent bankruptcies of WorldCom, US Air, Delta Airlines, General Motors, and numerous other firms that couldn't quite get it together over the past several years.

To further underscore this crucial point regarding people's false beliefs that it is events that catalyze the public's mood, let me recount a story told by Robert Prechter of an experience he had a few months after the Enron brouhaha hit the headlines. In April 2002 Prechter spoke at an Ivy-League symposium on the psychology of investing. His presentation followed the line of argument presented above, and he reports that several of the academics, psychiatrists, and investment professionals in the audience told him afterwards that the "mood-drives-events" argument caused them to reevaluate their views on social causality. But reevaluating views and actually changing them seem to be very different matters, since the reverse flow from event to mood was still fixed in the audience's mind as seen by what happened next.

A general discussion session took place at the end of the conference following Prechter's presentation, with the final question from the attendees to the speakers being, "The Enron scandal has deeply discouraged investors; when can we hope that this black cloud hanging over the stock market will go away?" The panel responded for 10 minutes or so, but no one questioned the basic assumption behind the question that it was an event (the Enron collapse) that *caused* discouragement in the investment community (a social mood). In short, despite Prechter's call to arms, the attendees to a man and woman retreated to the conventional wisdom that the Enron scandal caused investors to be discouraged. The socionomic position is just the opposite: Investor discouragement, as evidenced by a steadily declining stock market, gave rise to a climate in which an Enron-like scandal would almost surely break out, and that further such events would follow. And sure enough, they did.

But what do we actually *mean* by the term *social mood*? And how do we measure it? With the Enron situation as backdrop, we turn to these pivotal questions.

GETTING INTO THE MOOD

Most academics and continental philosophers seem to operate on the principle that "If I can't understand him, he must be a genius," and produce reams of documents supporting this claim. Fortunately, the public at large recognizes this sort of talk for what it is, namely, gobbledygook, and holds academics and other pundits to a different standard: "An idea has genuine content and value if and only if I *can* understand it." Following this precept, we will make use of the following commonsense notion as to what is meant by "mood": Mood is simply how an individual or a group feels about the future. Thus, *social mood* is how a particular group, community, population, or society feels about the future.

Social mood comes in two flavors: positive and negative. So if at a given time a group has a positive social mood, it means the group is optimistic and looks forward to the future; a negative social mood is just the opposite. In that case, the group is pessimistic and fearful of the future. Of course, there are degrees of optimism and pessimism, and not every person in a group feels the same. In fact, it is this heterogeneous aspect that gives rise to varying degrees of positive or negative social mood. But, in general, at any given time a group as a whole is on one side or the other of the positive–negative divide.

At this juncture the reader might well object to the black-and-white categorization of social mood as being either positive or negative with no shades of gray. While we've adopted the "one size fits all" policy of talking about *the* social mood of a population, fuzzing over the various subgroups making up the population, there is an important distinction to be made between social mood that's at a turning point and social mood that's turned and is still far from topping/bottoming out. To respect this difference, we will shift from black-and-white to grayscale, and allow for two different sorts of both positive and negative mood.

We take our cue for the differing types of mood from the categories introduced by Dominique Moïsi in his book *The Geopolitics of Emotion*. There Moïsi discusses three categories of social mood: hope, fear, and humiliation. Our notion of increasing social mood corresponds (approximately) to hope, while fear characterizes decreasing social mood. For Moïsi's term *humiliation*, we substitute *despair* to characterize the period when a negative social mood is bottoming out. Finally, to close the circle we introduce the term *hubris* to represent a peaking positive social mood. The table below summarizes this taxonomy of moods:

+ Mood rising	+ Mood peak	− Mood declining	− Mood bottom
Hope	Hubris	Fear	Despair

It's important to note here that while the mood is always positive or negative, it's also continually changing and moving from strong to weak and back again. This is the reason for the gradations just introduced. Most of the time the social mood is trending, in the sense that the mood tomorrow will be just like today—except a little better or a little worse, depending on whether the curent mood is positive or negative. Or in our new terminology, most of the time the mood is either hope or fear. There are even strong mathematical reasons for saying that this is the typical case, in the sense that if you pick a moment in time at random, there is probability 1 (which is *not* absolute certainty, at least not to a probabilist) that you're on a trending part of the social mood curve. In other words, the critical points where the mood is either peaking or bottoming form an infinitesmially small set in the collection of all moments in time. This fact, incidentally, accounts for why futurists of the trend-following persuasion are almost always right when they forecast that the current trend, whatever it is, will continue. But such forecasts are content free and next to useless, since you don't need to hire a futurist to tell you that tomorrow will be just a little better or a little worse than today. The real challenge rests in nailing the turning points. That information is real, not fool's, gold, and is something truly worth paying for.

In summary, then, the social mood is simply the beliefs a population holds about the future. Let me emphasize here yet once again that our concerns in this book are with the *social* mood, not the mood of the individuals making up the group. I've often heard people say things like, "Well, I feel this way and so do many of my friends" as an argument for why they believe the group must feel that way too. Things just don't work that way. Groups have an entirely different set of rules and dynamics by which they function, often totally at odds with the rules and dynamics by which every single member of the group operates—as an individual. Here's a depressingly familiar example.

Consider a group consisting of the fans of a soccer club. We regularly see reports of outbursts of violent behavior by such groups, actions like throwing trash and chairs onto the playing field, attacking the referee, and other not-so-random acts of violence perpetrated against supporters of the opposing team. When questioned, the people forming the group say they would never engage *as individuals* in the type of behavior they are happy to join-in with when they're acting as part of the group. In short, there is a group dynamic that emerges from the interaction of the individuals, a dynamic that simply does not exist at the level of the individuals themselves who make up the group.

Our concern in this book is with that group dynamic, not the thoughts, feelings, and beliefs of the individuals forming the group *taken as individuals*. Of course, those individual thoughts, feelings, and beliefs contribute to the group dynamic. No doubt about that. But they are not enough to create the group dynamic. For that, we need an interaction pattern among the individuals. Exactly how that interaction pattern serves to combine the feelings of the individuals into an emergent, collective group dynamic is very poorly understood, at least from a scientific point of view. But it is the essential element in generating the "social mood."

MADNESS, MEMES, AND MOODS

The spring of 2005 was a great time for crowd madness, at least if the crowd you were looking at was the one occupying corporate boardrooms across America. This period amply illustrates use of the term *madness* to mean "takeover mania," which was then striking every nook and cranny of this domain, as Proctor and Gamble announced that it was buying Gillette, SBC was trying to take over AT&T, Sprint acquired Nextel, Verizon tried to buy

MCI, and so it went, one firm after another engaging in an irrational fit of collective acquisition mania.

How can we call these "deals" an example of the madness of crowds? Simple: Just look at how the deals worked out. Although the deals were engineered by supposedly sophisticated CEOs armed with the best investment bankers and lawyers money can buy, these takeovers have all the earmarks of herding behavior of the very worst sort, at least for investors. The psychology of the situation is clear: When CEOs see a company in their industry buy another company, they become possessed by the dreadful vision that their industry is consolidating into a few giant firms. "What if my firm is left out?" they ask themselves. To prevent this worse-than-the-worst possible outcome, the CEOs immediately launch a frantic search for a company to buy with price being no object. What matters is sheer survival, or in most cases, perceived survival.

But a company can buy another company at such an inflated price that the stock of the parent firm never recovers. Of course, the acquisition makes the new, bigger company so big that it cannot be bought by another firm ("To Big to Fail"). But its shareholders are permanently damaged—while the shareholders of the acquired firm happily take the money and run. And this is not just armchair speculation (no pun intended) either. A recent analysis of the nine biggest takeovers in history, deals valued at $50 billion or more, showed that in seven of those nine cases the stock of the acquiring company never returned to the level where it was just prior to discussion of the purchase. But shareholders of the acquired firms received a generous premium to the price their stock was selling at just before the deal was publicly announced. Finance theorist Richard Roll has done an extensive study of this phenomenon, as reported in his article "The Hubris Hypothesis of Corporate Takeovers". There he reports that overconfidence on the part of the CEOs is the driving force of most mergers, despite the preponderance of evidence suggesting the merger is no bargain, at all, at least for the acquiring firm.

Despite the clarity of these facts even to the uninitiated, when merger fever strikes, rationality visiting the executive suite is about as popular a commodity as a visit from a tax inspector. The reptilian madness of the crowd always trumps the logical rationality of the "new brain." But, then, who would have ever thought otherwise? The puzzle crying out for attention here is how these waves of madness get started and spread through a group, especially one that's presumably as savvy and sophisticated as corporate CEOs, who one might suppose are receiving their not inconsiderable compensation packages for exercising prudent judgment of the sort that's needed to avoid exactly these sorts of negative consequences of group madness.

The basic idea underlying crowd madness is the same as the idea behind word-of-mouth advertising: spreading thoughts through a network of connections. Madness sets in when a new idea from the group emerges that is not held by any of the individuals in the group. In many ways this notion is analogous to the spread of a disease through a population, as we outlined briefly in the last chapter. An epidemic starts with some infected individual. That individual makes contact with many other persons, some of whom are susceptible to being infected. Those people, in turn, contact other susceptibles, and so it goes. The factors separating an epidemic from just a collection of sick people are the density of the web of connections, the susceptibility of the individuals to the disease and the virulence of the disease itself. So, for instance, a new strain of influenza is far more potentially dangerous than, say, Ebola fever, even though the latter is far more virulent. And that is exactly the point: Ebola kills its victims so fast that they have little chance to spread it to others even in a dense web of contacts. On the other hand, flu develops relatively slowly and is generally not fatal. So before you even know you have it you've contacted many others, some of whom are likely to contract the illness from you.

Translating this general notion of spreading ideas rather than bacteria and viruses, we have the basis for a kind of "thought contagion," whereby elementary ideas or snippets of cultural artifacts like popular songs or jingles can be transmitted from one brain to another via a mechanism not unlike the transmission of a flu virus. The evolutionary biologist Richard Dawkins christened such informational artifacts *memes*. They play the same role in informational structures in a population that genes play in the population's biological structure. Each individual in the population is continually trying to "infect" the brains of others with their own ideas and beliefs, just as at this very moment I'm trying to infect the brains of the readers of this book with this notion of a meme. Again, whether a particular meme like an iPod catches on or dies like the meme for a video phone call depends on the strength of the idea, how information about the idea moves around in the network of individuals involved, and the susceptibility of these individuals to catching the "idea virus."

As an example, if the population in question happened to be the citizens of North Korea the likelihood of an iPod epidemic would be small, since the susceptibility of any individual is low due to the lack of western currency to purchase the device. The connective structure may also be very unfavorable for an iPod epidemic in Pyongyang, where state control of the media and channels of communication ensures that very little information about this device will reach the population through the usual western-style modes of advertising, store displays, and the like. Before taking up the structure of information networks and their role in forming up a population's social mood, it's instructive to look at a couple of examples of this herding instinct in action.

MOODS AND MEMETICS IN ACTION

- *The John Kerry Postulate (*aka *The Self-Fulfilling Prophecy)*: In the run up to the 2004 US presidential election, all sorts of people were promising their vote to the challenger, Senator John Kerry. But despite the very bridgeable gap between Kerry and George W. Bush in the polls at the time, virtually none of these people seemed to think Kerry would actually win the election. I need not recount the end result. This phenomenon led to what political columnist John Vinocur termed the John Kerry Postulate: "Regardless of how narrow the margins in the polls look, your choice will lose if next to nobody expects him/her/it to win." Vinocur applied this Postulate to the French referendum in 2005 for approval of the European Union Constitution, noting that for two months prior to the vote the pollsters without exception were saying that the French voters would decline the Constitution. So virtually no one thought the voters would accept the Constitution—and, of course, they didn't. Vinocur observes that an Internet polling group called *Expression Publique* said that the "No" crowd were dug in as the majority, but that only 7% questioned the finding on the basis of their own experience. At this point the John Kerry Postulate kicked in, suggesting a rather bleak prospect for approval of the Constitution.

  This is an excellent example of how the prevailing social mood biased a collective event, in this instance the outcome of a popular vote. Despite the herculean efforts of the French government to smoke out a "Yes" decision from the voters, no one believed in France's role as a major player in the EU. So the vote went against approval of the Constitution.

- *Spread of Change at Nortel*: Nortel is a leading data and telecom supplier and manufacturer employing many thousands of people worldwide.

In October 1997 the company sponsored an event offering participants in training and organizational development within the firm the opportunity to learn from each other's successes (and, presumably, failures too). The agenda of the meeting involved hands-on experience with a program called *Tipping Point*, which was a simulation of the organizational change process. The simulator was modeled on the way a disease spreads in a population, using that process as an analogy for the way ideas spread throughout an organization. Within the simulation, proponents of new ideas are analogous to people infected with a disease. So their enthusiasm for whatever idea they are promoting is literally . . . infectious!

By word of mouth, advocates spread the news about the benefits of the new idea. The rate at which the news traveled depended on the number of contacts between advocates of the idea and those who were neutral or negative toward it. It was also dependent on external factors, such as rewards and recognition for adopting the innovation, infrastructure support for whatever new processes the idea required for its implementation, and support of influential leaders in the organization. As the number of "infectives" increased within the organization, the idea eventually reached a critical mass causing the entire organization to flip sharply in support of the new initiative.

At the Nortel conference, 20 teams competed against each other to design the best strategy for increasing the number of advocates of a given innovation, while at the same time limiting the cost of promoting the idea. The most important lessons the participants drew from this exercise were that the best advocates for innovations are people who accept and apply change, that connections between advocates of an idea and others in the organization are essential, and finally the pivotal role played by higher management in Nortel serving to shape the kind of behaviors they want to see.

- *Wisdom of Crowds*: Journalist James Surowiecki has noted that the collective wisdom of a large group of people making independent guesses about the outcome of an event like an election may well shed more light on the actual outcome than listening to any individual "expert." In other words, the mood of the group can trump any individual opinion or sentiment. For instance, if you had paid attention to journalists during the 2004 American presidential campaign, you would have read about a tight race with daily ups and downs favoring one candidate or the other. Just prior to the election, most pundits were saying the race was simply too close to call in the major battleground states. But the wagering on *Intrade*, an Internet betting site, painted a very different picture. Bettors were calmly putting their money on a Bush victory. Even more remarkably, the weekend before the election bettors called the winner in *every one* of the 50 states!

Surowiecki's book on collective wisdom recounts many examples of when the crowd gives better answers to a question than any single advisor or expert. For instance, he tells about the dramatic event in May 1968 when the nuclear submarine *USS Scorpion* sank on its way back to its home port at Newport News, Virginia. The US Navy had an idea of the general area in which the sub went down, but due to communication blackouts designed to hide the position of nuclear submarines, as well as technical limitations on communication from the deep, the *Scorpion* was not able to broadcast its actual position at the time it sank. So all the information the Navy had was that the sub went down in an area 20 miles wide and thousands of feet deep. To pinpoint the precise location of the sub, naval officer John Cravens hit on the idea of gathering a group of diverse experts and asking them for their best guesses as to why the sub went down, what its rate of descent was as it sank, the angle of its dive to the bottom, and other information of this type. Craven took all these guesses and processed them via some mathematical wizardry to generate the group's overall best estimate as to where the sub was located on the ocean floor. It turned out that the Navy found the *Scorpion less than 220 yards* from where the group predicted it would be located. Yet not one individual in the group had picked that specific location—or any location as close to the true location as that estimated by the group as a whole.

Of course, crowds can be disastrously wrong in their collective judgments too. Witness the Dutch tulip mania of the 1630s, the Crusades, or the bowling stock bubble of the 1950s, in which futurologists assumed every man, woman, and child in the United States would bowl for two hours a week, 52 weeks of the year. Among the many factors separating the glory of the *Scorpion* identification from the ignominy of tulip mania are the need for diversity, and most importantly, independence of the individual opinions (no "group think"), and mechanisms for aggregating the individual estimates into a collective judgment. The Internet markets, such as the Iowa Electronic Market for elections, fit the bill perfectly in this latter regard, as does the Hollywood Stock Exchange for betting on the outcome of box-office returns for new films.

How, *exactly*, are the thoughts and especially feelings of individuals in a population herded together into a kind of "madness of crowds" that constitutes what we see as the social mood? No one really understands in detail how this happens. But in recent years complexity scientists have developed theories for describing the process. Not surprisingly, these theories involve how people are connected in a network and the way information percolates throughout a network to move feelings, opinions, and thoughts from one brain to another.

THE CONNECTIONS THAT COUNT

It's not very often that a major intellectual advance is first announced in a popular magazine. Revolutionary ideas usually first see the light of day in obscure academic journals, and only much later, often very much later (or never), make their way to the pages of the popular press. A notable exception occurred in the general-circulation magazine *Psychology Today*, when in 1967 it published an article by Stanley Milgram, probably the greatest social psychologist of the twentieth century. In this piece, Milgram described the results of an experiment he carried out to test how many intermediate acquaintances, or links, would be needed to connect two randomly-chosen people in the United States.

To explore this question, Milgram first chose two target individuals, a divinity student living in Sharon, Massachusetts, outside Boston, and a stock broker living in Boston itself. He then prepared 160 letters, which he divided into two lots and sent to randomly selected people in Wichita, Kansas, and Omaha, Nebraska, towns chosen because in Milgram's view they seemed to exemplify Gertrude Stein's withering putdown of Oakland, California: There's no "there" there. He asked the recipient of each letter to send it on to an acquaintance they thought would help promote the goal of the letter eventually reaching its target in Massachusetts. So, for instance, if one of the people in Wichita or Omaha happened to actually know the target, then there was no intermediate link. On the other hand, if there were 100 people in the chain from Wichita/Omaha to the target, then the connection involved 100 links. What were the results? How many of the 160 letters actually made it to their target? And how many intermediate acquaintances were necessary, on the average, to achieve this connection?

Prior to the experiment, Milgram had asked many intelligent, knowledgeable people what they thought the answers would be. Most guessed that there would be many links between the source in the Midwest and the target in Massachusetts. One person even thought it might involve 100 intermediate acquaintances to complete the chain. After mailing the letters to the random sources, Milgram waited nervously to see if *any* of the letters would eventually reach their target. To his surprise, within a few days the first letter came back having gone through only two intermediate links. This was simply beginner's luck, though, as in the end just 42 of the 160 letters finally reached their target. The greatest surprise, however, was that of these 42 letters, the average number of intermediate links was just $5\frac{1}{2}$—a number vastly smaller than Milgram anticipated. Playwright John Guare rounded this number up to 6, and later titled his 1991 Broadway hit, *Six Degrees of Separation*, enshrining this term into everyday language.

Although the experimental conditions underpinning Milgram's work have been questioned in a number of ways, the overall conclusion still seems to hold. Even though the world consists of 6 billion or so people, an astonishingly small number of links are needed to connect any two individuals. No one is more than a few phone calls or handshakes or mouse clicks away from anyone else. This empirical observation has led to the term "small worlds" as a catch phrase for this type of network connectivity. Since the structure of a network is at the heart of how information moves from one person to another to create an overall social mood, we need to understand as much as possible about the way the pieces of a network can be connected, and how those connections allow the flow of information from one part of the system to another. Let's take a peek at what network theorists have discovered about these all-important matters.

The usual procedure for studying the connective structure of a group is to represent each member of the group by a point (node) and then draw a line (arc) between any two people who are directly connected by sharing a communication link, friendship, or some other direct mode of interaction. This gives rise to what mathematician's call a "graph," which is simply a set of nodes and the arcs linking various pairs of nodes. Graphs that are *clustered* have many nodes that tend to form closely-knit groups with many arcs connecting people in the group, but very few arcs leading out of the group. On the other hand, a graph whose arcs are selected at random has no such clustering structure. Figure 2.2 shows two graphs, each with 12 nodes and 18 arcs, displaying these two types of connective patterns.

Milgram's problem was to discover the smallest number of links it takes to reach any other node from a given node in the graph. The average of all such shortest paths in a graph is called its *characteristic path length (CPL)*,

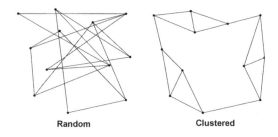

Figure 2.2   A random network and one that is clustered.

which then is a measure of the degree of separation in the network. For the random graph in Figure 2.2 the CPL is 1.95, while it is 2.45 for the clustered network. Milgram's studies indicate that even when the number of nodes is several hundred million, or possibly even a few billion, the CPL only goes up to around 6. But neither the random network nor the fully clustered one really reflects the structure of the networks of everyday life. For instance, if the US airline network were random, on average there would be the same number of flights going from Chicago to Los Angeles as from Chicago to Canton, Ohio, or Tuscaloosa, Alabama. But if it were clustered, it would take a large number of flights to travel from one city to any other. Yet we all know that it seldom takes more than two or three flights to reach just about any town in the world from anyplace else. So there must be some other structures in real networks that are not present in these two extremes. Milgram's work suggests what's missing.

In another version of his experiment, Milgram sent 60 letters to various people in Wichita. In this case the participants were only allowed to pass the letters by hand to personal acquaintances they thought might be able to reach the divinity student in Massachusetts. Fifty people responded to the challenge—but only three letters eventually made their way to the target. In this and similarly-related experiments, Milgram noticed a kind of "funneling effect" whereby most of the letters being forwarded were by a small number of "hubs" who were significantly more well-connected in the network than the average person. Even in the original study in which only three chains were completed, Milgram noted that "two of the three completed chains went through the same people."

This experiment leads to the idea that real networks have "shortcuts"—a few people that connect clusters that would otherwise be very far from each other. Alternately, there may be a few people who have an extremely large number of acquaintances in the network ("connectors") who serve as hubs

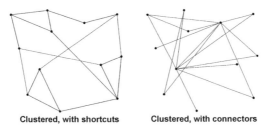

Clustered, with shortcuts   Clustered, with connectors

Figure 2.3   A clustered network with a shortcut and one with connectors.

through which a large volume of interactions flow as with the airline networks. Examples of these two types of graphs are displayed in Figure 2.3. Interestingly, the characteristic path length for the so-called "small worlds" network, a clustered network with shortcuts, shown here is 2.03, virtually the same as for a random network. But the clustering structure is much higher in the small-worlds case. For the network with connectors the CPL is 1.88, a bit smaller than for the other graphs but not significantly less. Again in this latter case the clustering is much greater than for the random network.

In his book *Small Worlds*, Duncan Watts shows how small-worlds connectivity structure can be constructed very easily from the fully clustered graph of Figure 2.2. He also points out that it takes only a small number of shortcuts in a highly structured graph to turn it into a small-worlds network. But one thing the small-worlds structure does not have is hubs. These arise in a information structure when two factors are present: (1) the network grows by adding new members to the population, and (2) the new members connect in a preferential fashion linking most often to those people who already have a lot of links. The consequence of these two simple conditions is that the network becomes what Albert-Lázló Barabasi terms a *scale-free network*, since there is no typical number of nodes per arc. These types of networks also display what's termed a *power-law* structure, which in everyday language means that there are a few nodes having an extremely large number of connections (the connectors) and lots of nodes that have very few connections. This is exactly what we see in structures like the Internet and interactions among cellular metabolites. In both cases there are a few nodes—web sites like Google or particular molecules—that are very large (many connections), while most web sites and molecules are relatively insignificant (small number of connections).

The small-worlds networks have a kind of "egalitarian" structure, in which there is a distribution of people around the world such that the vast majority of the population is weakly connected to each other, while a few relatively "powerful persons" like the Pope, a famous athlete, or Bill Gates have a huge number of connections. To see the power of just a few random links in an otherwise clustered network, science writer Mark Buchanan has estimated that in a clustered network where each person is connected only to his or her nearest 50 neighbors, there are 60 million degrees of separation needed just to go halfway around the world's population. But if there are just 3 random links out of each 10,000 links, then the degree of separation falls from 60 million to only 5!

By way of contrast, a network with connectors is not egalitarian at all. It is "aristocratic," with just a few people having an enormous influence on the movement of information from one part of the network to another. For instance, estimates of path lengths on the World Wide Web suggest that there are between 4 and 10 degrees of separation between any two web sites. Thus, it should not take more than 10 mouse clicks, on the average, to go from a randomly chosen site to any other. But it has also been suggested that every time the number of links in the Web doubles, the number of nodes possessing the previous 4 to 10 degrees of separation is reduced by a factor of five. Consequently, the "rich" nodes get richer and the poor relatively poorer, where here the "rich" nodes are those sites like *Google*, *Microsoft*, *iTunes*, and *amazon.com* that are the connectors or hubs that dominate the Web.

What does all this have to do with the kinds of social networks we're concerned with in this book? Which of these types of graphs is the most likely candidate for representing the structure by which information about individual opinions and feelings about the future pass through a population? The simple answer is, I don't know. To the best of my knowledge the question has never been explicitly studied using real networks and real data. But it's not much of a gamble to speculate that the best candidate is the last one, the clustered network with connectors. Establishing (or refuting) this conjecture is a very small piece of the research program given in Appendix B that I feel is needed to put the hypotheses presented here on solid scientific ground. Let's conclude our study of networks and social mood by examining the role connectors play in "flipping" the social mood from one extreme to another in a rather short period of time.

Probably the most over-hyped method for losing weight that's been foisted off onto a desperate and obscenely overweight public is the so-called

Atkins Diet, named after a New York City physician, Dr. Robert Atkins. According to Malcolm Gladwell, author of the best-selling volume *The Tipping Point*, a large part of the appeal—and publicity—behind this diet is that it reframes the question of dieting from thinking about weight loss in terms of avoiding calories and fat to thinking about it as avoiding carbohydrates. Such a shift totally changes the way people look at dieting. This is a particularly transparent example of the much more important fact that if you want to know if some event suggests the close proximity of a turning point, be it in dieting or in world affairs, ask yourself: Does that event cause people to reframe an issue? Publication of the Atkin diet signaled a turning point in the world of dieting.

In his book, Gladwell identifies three elements necessary for an unstable situation to flip from one state to another. He calls them *The Law of the Few*, *The Stickiness Factor*, and *The Power of Context*. The first asserts that only certain types of individuals can serve effectively as transmitters of information. He terms these the *Connectors*, the *Mavens*, and the *Salesmen*. We have already seen the first type. They are people who know everyone. The Mavens are a variant of the Connectors: They don't know everyone. Instead, they know everything. A football Maven, for instance, knows everything about the world of football, who's hot, who's not, who's injured, what team is ineffective against what other team, and so on. Finally, the Salesman knows how to do a good job of communicating information to others. Gladwell cites Paul Revere as the quintessential example of all three, since he knew what the British were up to (Maven), knew all the right people to tell (Connector), and was able to spin a good story about the situation (Salesman). Thus, a model aimed at understanding shifting social moods would necessarily have to incorporate all of these types of "transmitters" as well as a much larger number of people who cannot serve any of these functions very well. We might term these people The Herd. Here are a couple of examples illustrating different possibilities.

A lot of computer afcionados were overjoyed when Linus Torvald created the open-source operating system Linux in the early 1990s. Linux was trumpeted to the world as an alternative to Windows by the anti-Microsoft crowd, with some degree of justification given the seemingly unending series of reports of bugs and security holes in each successive version of Windows. But what's happened to Linux? How much market penetration has it actually managed to achieve in the decade or so that it's been freely available? No one can really say. But even a generous estimate is around 29 million as of spring 2005. This sounds like a lot. But Google estimates this is only about 1% of the machines that query its search engine. I'll leave it to the reader to guess whose operating system is running on the vast majority of the remaining 99%. So why hasn't Linux reached "escape velocity" and soared off into the operating system stratosphere? According to most knowledgeable sources, Linux is technically far superior to the competition, costs nothing, and has lots of application programs that run on it. Why hasn't Linux yet "flipped" to become the operating system of choice?

According to John C. Dvorak, a keen observer of the computer scene and regular columnist for *PC Magazine*, the reason for Linux's failure to reach escape velocity is simply the absence of a critical,"must-have," end-user application that runs on Linux and Linux alone. Dvorak points out that earlier operating systems that succeeded, at least for awhile, all had killer apps like VisiCalc for the Apple II. Those operating systems that were technically outstanding, such as the NorthStar in the 1970s, had no killer app and died on the vine. Linux appears to be in a situation where it can serve a niche market due to its Apache Web server software. But this is simply not enough to create a tipping point. The only thing that will do that, according to Dvorak, is the Linux-exclusive, must-have application. No amount of fancy user interfaces, technical virtuosity, or other whistles and bells will suffice. But the inherent conflict in having an open source code, together with Microsoft's dominance and stifling of industry innovation at every turn, leads to the sad conclusion that there is not going to be any such killer app. Neither Linux nor the Mac is going to achieve product escape velocity and flip to become the operating system of choice for the typical end-user. In this sense, both are condemned to a life in the shadows of the computer world, serving niche markets of specialists, connoisseurs, and computer fanatics. In terms of information contagion and idea viruses, Linux just didn't start an epidemic in the world outside that of computer hackers, geeks, and those already in the know. If Dvorak's analysis is to be believed, it never will.

A happier story of a phenomenon that not only went beyond the tipping point, but actually threatens to change Gladwell's analysis of what is needed for such a feat, is the process of information exchange on the Internet via web logs ("blogs").

A couple of years ago, John Hiler published an article on the Internet, sent notice of it to a friend who published a widely-read blog for high-tech news, and began thinking about other things. To his astonishment, within a week more than 30,000 visitors came to his site to read the article. In analyzing this unbelievable (to him) reaction to his modest posting, Hiler came to some conclusions that have direct bearing on our question as to how individual moods can "take off" to tip the prevailing mood of the population as a whole. Here's a brief summary of what he discovered.

First of all, Hiler noted that every epidemic needs both a host and a virus. He claims that the virus is simply the hyperlink, while the blogs are the host. When it comes to ideas as viruses, there can be no better vehicle for the virus to be than a link. Each link represents a web page holding out the promise of fascinating new material to lure the curious browser. Moreover, a link can copy itself from one blog to another at nearly the speed of light, generally by human-directed intervention.

In Gladwell's taxonomy outlined earlier, there's a clear distinction between the Connectors who know people and the Mavens who know facts. Hiler correctly points out that this distinction is blurred more than ever by the information transfer process on the Internet. The reason is pretty simple. Offline Connectors need to be gregarious sorts of people who move in many circles and are always eager to meet new acquaintances. Online Connectors need have none of these qualities. It's a lot easier to blog a posting than it is to make a presentation or go to a convention. So suddenly blogging Mavens are as important as Connectors as far as being a hub is concerned. Why go through the intermediary of a Connector when you can get the real scoop directly from the Maven? In short, eliminate the middle man.

To summarize how his article spread on the Net, Hiler identifies five steps:

1. Someone (an "expert" or "content Maven") creates a document and puts it online. This generates the link.
2. A "link Maven" sees the link and blogs it on his or her site.
3. A Connector (or a link Maven with Connector-like traffic to his or her site) finds the link and blogs it to his or her site.
4. The link achieves "escape velocity" in the blogging community.
5. The link then gains escape velocity *outside* this community.

This process turns the traditional view of advertising and marketing (i.e., viral marketing) totally upside down. As Gladwell himself notes in the Foreword to Seth Godin's illuminating volume *Unleashing the Ideavirus*,

Advertisers spent the better part of the 20th century trying to control and measure and manipulate the spread of information—to count the number of eyes and ears that they could reach with a single message.

But this [ideavirus] notion says that the most successful ideas are those that spread and grow because of the customer's relationship to other customers—not the marketer's to the customer.

In short, it is "word-of-email" advertising from friends and acquaintances that transmits the message, not the unwanted intrusion of a marketer's message. So in today's world of high-speed, highly-connected communication networks, if you understand web Mavens and Connectors you can spread your ideas—including your mood(s)—beyond simply web logs to the population at large.

We now want to leverage up our working definition of social mood into the beginning of an actual theory of how collective events are conditioned by the mood of a population. For this, it's necessary to have some means to actually measure the social mood in a population at any given point in time. By taking the stock market averages as a surrogate measure for the overall social mood, it follows that the predictive power of a social mood hinges upon our ability to ferret out patterns in stock prices movements and infer from these patterns what types of events are more or less likely to unfold over different time horizons. Let's now compare various alternatives to stock price movements as a social mood measuring stick.

MOOD METERS

FutureMe.org is a web site where you can schedule transmission of an email to yourself to arrive at any time between now and 30 years into the future. Researchers Johan Bolen of the Los Alamos National Laboratory in New Mexico and Alberto Pepe of UCLA applied a mood-rating system to the text of over 10,000 such emails sent in 2006 to characterize the senders beliefs and feelings about the future.

The researchers discovered that emails scheduled for delivery in 2007–2012 were *significantly* more pessimistic (read: negative) about the future than messages targeted for delivery in the following 6-year period 2013–2018. This raises the question whether this type of social mood meter could have been used to forecast the global economic meltdown that began in 2007. In other words, do the online searches we do today provide a way of measuring the social mood that gives rise to events tomorrow?

As an illustration of this possibility, the Seattle firm Veratect, Inc. used a proprietary computer algorithm to monitor online chat rooms, blogs, and offline information sources looking for hints of civil unrest and outbreak of global diseases. The idea behind this work is that such data sources contain information that officials can use to respond more quickly and effectively to problems like the spring 2009 swine flu outbreak. According to the firm, their work alerted readers of their subscription-only newletter to a potentially severe outbreak of influenza before swine flu hit headlines all over the world. More specifically, Veratect claims that it posted a report on April 6 outlining an unusual number of respiratory cases in Mexico, and then followed up that report on April 16 with an alert to the Center for Disease Control in Atlanta. It's worth noting that search engine giant Google reported similar types of results from analyzing search requests in Mexico during the same period. Of course, these are all just data that still must be classified, positive or negative, in order to get a measure of how the data points to people's feelings about the future.

What we seek is a kind of "sociometer," an instrument that plays much the same role in measuring the "temperature" of a population's mood as a thermometer plays in measuring the temperature in a room or in your body. It's important to note that neither measuring instrument, the sociometer or the thermometer, surveys the entire population, be it the views of each individual in a group or the motion of each and every molecule in a room. Both take a representative sample of the population and infer a "heat content" for the entire population from that sample. One might say that if the reading of the sociometer is "hot," then the mood of the group is positive, while if it's "cold" the mood has swung to the negative.

Our question about Internet data-mining procedures is whether these methods can serve to generate a good measure of social mood in different populations? Skeptics wonder whether companies claiming to do this are actually able to ferret out meaningful signals from the snowstorm of data sources, or whether they're just good at finding patterns in the data after the fact. For our purposes, though, what matters is not the pattern-recognition aspect but rather whether this type of data-mining gives rise to a reliable way of measuring social mood. A method for interpreting how that mood will be translated into actual events is a kind of add-on extra. So we'll keep this possibility of the Internet in mind as we look at a few other candidates for a good sociometer.

PUBLIC-OPINION SURVEYS AND QUESTIONNAIRES

The most obvious candidate for measuring the social mood is just to ask people how they feel about the future. In essence, this is what surveys and public-opinion polls do in order to create the endless stream of media reports forecasting the outcome of elections, football games, Oscar winners, presidential popularity, and other questions of topical concern. Of course, pollsters employ a huge variety of statistical tricks and sampling procedures designed to identify a small subset of the population that is representative of the population as a whole. Nevertheless, comparing the results of these polls to the actual outcome of events doesn't inspire much confidence that polls are a very useful measure of social mood (or anything else, for that matter).

Part of the problem with using surveys as a sociometer is that they reflect only what people *say* they think or feel. And people might—and do—say almost anything. But when it comes to crunch time in the election booth or in a betting pool at the local sports bar, people often *act* in an entirely different manner from what they say. This phenomenon surfaces regularly in surveys surrounding the outcome of elections, as pollsters continually misread the mood of the population and forecast results wildly at variance with what actually takes place. The most famous failure of this sort is the 1936 *Literary Digest* election poll, in which the magazine sampled 2.27 million owners of telephones and automobiles concluding that Franklin Roosevelt would lose the election to Republican Alfred Landon. He didn't; in fact, it wasn't even close. The problem was that a lot of 1936 depression-era Roosevelt voters didn't own telephones or automobiles so never got the chance to voice their opinions. Modern pollsters now know enough to avoid these kinds of obvious mistakes. But the gap between words and actions still remains.

There are other objections to the use of surveys and questionnaires as sociometers, too. For instance, the data are not readily available on all timescales, there is not a huge database of historical data to draw upon, and the measurements are filled with inaccuracies of many types. Moreover, behavioral psychologists have shown time and again that the results of a poll are strongly conditioned by the form in which the questions are framed by the pollsters. Taken together, these objections force us to look farther afield for a sociometer of choice. In particular, our preference is to identify a social mood meter based upon actions rather than words. Here are a couple of candidates.

ANNUAL BIRTHS

When people look forward to the future and think tomorrow is likely to be better than today, they are much more inclined to have children than when they feel the future is dangerously uncertain. This seems pretty self-evident and is actually borne out by the historical data on births. Now how can I make this claim, since to compare births with the historical record of optimism/pessimism about the future involves being able to measure the views people have about the future, i.e., the social mood? Mostly, such studies have employed current economic conditions like employment rates, averge worker earnings levels, and the like to characterize the mood, arguing that when people have jobs and are seeing increased earnings they are more optimistic about the future than when they're unemployed and/or experiencing a decline in purchasing power from their paychecks.

As a piece of recent evidence for the relationship between annual births and social mood, the US National Center for Health Statistics reports that not only are there fewer jobs in the recession, there are fewer births too. US births fell in 2008, the first year of the current recession, ending a "baby boomlet" that had been in place since the start of the decade. The agency also reported that 2007 was a year that saw more babies born in the United States than in any other year in the country's history. Experts note that the Great Depression and all recessions since then were accompanied by a reduction in births. Digging just a bit deeper into the report, we find that the worst-hit states were Florida and California, the two states most affected by the downturn in housing prices.

Unfortunately, while annual births are strongly correlated with measures of positive and negative social mood like stock price movements, the birth record is also not a very useful sociometer for a lot of reasons. First of all, by their very nature annual births are recorded only . . . annually. So it's not possible to use these data to see anything other than events that unfold on a timescale that's yearly or longer. Also, the data on births are not entirely straightforward to obtain, at least in many parts of the world. Finally, the data are conditioned by changing social mores and shifts in attitudes toward having families as opposed to having a career. In this latter connection, we need only look at the plummeting birth rates in western Europe over the past couple of decades as runaway economic growth and increasingly positive outlooks about career prospects, especially for women, have reduced birth rates to below replacement level.

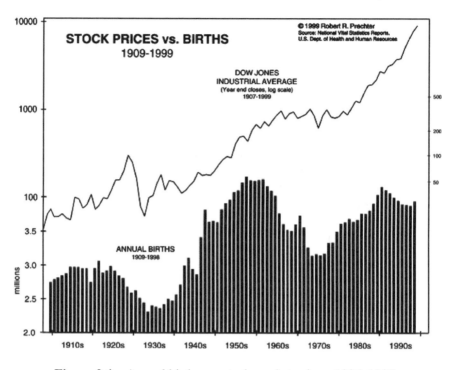

Figure 2.4   Annual births vs. stock market prices, 1926–1998.

A recent story on the Internet carried the headline "Recession Means Fewer Babies: US Births Fell 2 Percent." Later in the story, we find the statement "The downturn in the economy *best* (my italics) explains the drop in maternity, some experts believe." Well, there's hardly any idea so bizarre that you can't find "some experts" who won't believe it. But just in case you might be thinking along the lines of "some experts" that a low number of births are a *consequence* of bad economic conditions, not a precursor, have a look at the picture in Figure 2.4. Note especially the period in the late 1920s, where annual births were already falling long before the crash of 1929. So it wasn't the crash that caused people to be uncomfortable with starting a family. They were already uncomfortable.

INTERNET MARKETS AND SEARCH TRENDS

Earlier we mentioned that the collective wisdom of an Internet market like *Intrade* can often reflect the collective mood of the populace, a mood that then translates into actions like the outcome of a papal or presidential election. Even more importantly, measures such as movement in the odds on a particular event in a market like *Intrade* serve admirably as a way of characterizing and measuring the fluctuations of the social mood in a given population. As the notion of social mood and its measurement is the central pillar upon which our theory of social action and behavior rests, we should look a lot harder at this argument claiming that an Internet market movement serves as a useful "sociometer" for measuring the group mood. But there are problems.

The first problem in using an Internet market to measure the mood of a population is the question, Which population? The famous *New Yorker* cartoon showing two dogs, one at a computer terminal and one sitting on the floor, illustrates the point. The dog at the terminal looks to the other and says, "On the Internet, no one knows you're a dog." And so it is. On an Internet market you have no idea who the other bettors are, where they live, or in general, what actual population they belong to (other than the population of that particular Internet betting site). So what *social* mood is it that the betting site is actually measuring? At least with the Standard and Poors 500 Index (S&P 500), you can believe that it mostly represents American interests, while a movement on the Straits-Times Index in Singapore reflects bets being made primarily by Singaporeans, or at least people from Southeast Asia. But on the Internet, no one knows if you're an American or a Singaporean or . . . a dog.

The second barrier to using changes in Internet market odds as a sociometer is that there is very little historical data available to calibrate such market movements with actual events. These markets are simply too new to have amassed any substantive body of past data to mine.

These are the two most serious obstacles to the use of Internet trading markets as our sociometer of choice. The second barrier might disappear over the course of time. But the anonymity of the Internet is still likely to make it very difficult to overcome the first one.

In the last year or so a number of firms have sprung up to provide what's coming to be termed *sentiment analysis*. This involves mining the vast sea of data on the Internet in an attempt to turn opinions into hard data reflecting the collective consciousness of Internet users. The reason this work holds out promise as a commercial undertaking is that online opinion has become a sort of virtual currency that can make or break a product in the marketplace.

Here's an illustration of how the basic idea underlying the services offered by firms like Scout Labs, Jodange, and others in the sentiment analysis business can be used in practice. In May 2009, the ticket marketplace StubHub identified a rising negative sentiment on blogs after a baseball game between the New York Yankees and the Boston Red Sox was delayed by rain. It seems that stadium executives had mistakenly told many fans that the game had been cancelled, and StubHub was denying fans refunds on the grounds that the game was actually played. But after seeing virtual storm clouds brewing up in the blogosphere, the company offered discounts to the affected fans. This incident has now caused StubHub to re-think its bad-weather policy. As John Whelan, StubHub's director of customer services, put it, "This is a canary in a coal mine for us."

A closely related possibility for a sociometer is something like Google Trends, which gives a more-or-less real time indication of what search terms are "hot" in a given region of the world. While still suffering from the lack of historical data, this indicator is a stronger candidate than the Internet markets because it reflects things that people in a given geographic region are concerned about at a particular point in time. Of course, it's still a huge step from seeing a surge of searches in Mexico using words like *phlegm, flu, fever*, and *pigs* to infer something like a swine flu epidemic. Nevertheless, textual data-mining tools are improving daily and such searches might well turn into a usable and useful sociometer. But not quite yet.

Before leaving this category, let me present an exercise in measuring social mood via this type of "text mining" cum interview that holds considerable promise as a sociometer. This is something called a *Semiotic Image Wall*. I'll describe it within the context of a study done in 2008 together with my colleague, Leena Ilmola, for the government of Singapore, which aimed at characterizing the social mood in that city-state.

The first step is showing (via the Internet) the images depicted in Figure 2.5 to a sample of the population, in this case a few hundred Singaporeans who responded to an invitation issued by the government to participate. Respondents were asked to choose the image that best displayed their view of Singapore today, as well as the image that best fit their view of Singapore's future. The respondents were also asked to describe *why* they selected the images they did. For instance, one participant chose the ducks image at the lower right as descriptive of his (or her) view of Singapore today, saying "The orderliness of the ducks, with the same mould (sic!) from the same 'factory'." The point of the exercise was not to study the images that were chosen, though. Rather, it was to dig into the verbal responses themselves. The images were simply vehicles to elicit the verbal responses, which were the actual focus of analysis.

Figure 2.5 A semiotic image wall.

Using proprietary software for extracting semantic content and connections from the various responses, structures linking various words and concepts were then generated. We used these semantic webs to identify words and phrases suggesting positive or negative social mood in the population. A similar analysis was carried out on more than 90,000 articles over a 20-year period taken from the *Straits-Times*, the main newspaper in Singapore, looking for words and phrases suggestive of the population's view of the present and the future.

This analysis led to a division of the population into four groups according to whether an individual was positive/negative about today and tomorrow. The four groups each play a very different role in society. Those uniformly positive and thus happy with today andlooking forward

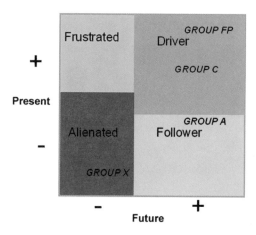

Figure 2.6   Social division in Singapore society.

to tomorrow are the Drivers of society, while those happy with today but pessimistic about tomorrow are the Frustrated group, those most likely to emigrate. On the other side of the fence we have people who are unhappy with the current situation but optimistic about the future, the Followers, and finally the Alienated, those who are unhappy today and think tomorrow will be at least as bad. Figure 2.6 shows how the population in Singapore in 2008 broke down into these four categories. The Drivers dominated the picture, followed by the Alienated, suggesting a bipolar division in society between the Drivers and the Alienated, with the Frustrated and the Followers being more like observers of this struggle.

Using this type of breakdown, we can identify those "zones" where government and social institutions can breathe easy (the "free zone") versus times when they should be concerned (the "alert zone"). This split is shown in Figure 2.7.

The take-home message from this exercise in measuring social mood is that interviews evoking people's feelings about the present and the future, coupled with technological tools for "mining" those interviews instead of asking directly how people feel, can lead to very useful conclusions as to how a society is divided at a given point in time, as well as how it is likely to evolve over a relatively short-term horizon.

On the negative side, the process of semantic textual analysis is time-consuming and labor intensive. Also, it does not offer an obvious procedure for projecting into longer-term futures. So we must continue the search for

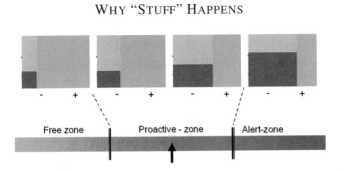

Figure 2.7   Zones of social mood.

a sociometer that combines the desirable features of the semiotic image wall with some easily applied, natural procedure to forecast the mood in the future from where it is today. Enter Wall Street!

FINANCIAL MARKET INDEXES

A few years back, journalist Kenneth Chang stated, "A market is the combined behavior of thousands of people responding to information, misinformation and whim." This pithy observation captures admirably the spirit of why the movement of a financial market index, such as the Dow Jones Industrial Average (DJIA), serves so well as a sociometer. As Robert Prechter states, "The stock market is far more significant to the human condition than it appears to casual observers and even to those who make their living by it. The level of aggregate stock prices is a direct and immediate measure of the popular valuation of man's total productive capability."

In other words, a stock market is in effect a huge casino where people place bets about the future on all timescales. As bets come in, the market assimilates them in nearly real time into a single number: a change of price in the various stocks that investors and speculators are betting on. A market index like the DJIA or the S&P 500 index simply measures price change in a representative sample of the major stocks being traded on a given exchange. Such indexes serve as a "thermometer" of the overall market. So if investors are optimistic about the future on a given timescale, they tend to buy stocks with that time horizon in mind and the index rises. Conversely, when they are pessimistic about the future, the majority of investors sell and the index declines. Thus, as Prechter puts it, the stock market is "literally a drawing of how the scales of mood are tipping. A decline indicates an increasing 'negative' mood on balance, and an advance indicates an increasing 'positive' mood on balance." It's that simple.

Let's have a look at why a financial market index is such a sterling candidate for our sociometer of choice. We'll turn to some of the possible objections afterwards.

- *Data availability*: In most industrialized countries, financial markets have been around for a century or more. As a result, there is a *lot* of data available on *all* timescales.

- *Ease of gathering data*: In contrast to many of the candidate sociometers discussed earlier, obtaining the financial data is pretty straightforward. At the simplest level, you get it from your daily newspaper. For finer-resolution data, numerous data sources exist to provide weekly, daily, hourly, or tick-by-tick data, sometimes even for free. So there is no problem at all in getting whatever data you want on whatever timescale you choose.

- *Clean data*: In experimental science, one of the most vexing problems is missing or corrupted/contaminated observations. There's no problem in this regard with financial data. It's about as "clean" as any data you'll find anywhere about anything.

- *Data analysis tools*: There is probably no set of data that has been analyzed and re-analyzed by more tools and procedures than financial data. Bookstore shelves sag under the weight of volumes that have been published outlining procedures for projecting financial data into the future. And the Internet is filled with adverts offering a bewildering array of tools and/or advice for this purpose. So, again, absolutely anyone can analyze the patterns in these data to as fine a degree as they might wish.

There is no candidate sociometer that I can think of that can even begin to touch financial data insofar as the above desiderata is concerned. But when I've offered this argument in lectures, members of the audience so routinely raise the same concerns that it's worth taking a moment right now to address the most frequent ones. And by far the most common reservation centers around the fact that stock prices are determined by actions of investors. But investors making direct decisions on buying and selling represent only a minuscule fraction of the total population. So how can a financial market index represent the mood of the *entire* population? Let me lay to rest this tiresome—but very common—complaint once and for all.

The distilled essence of this objection seems to be that a social mood meter must *directly* take into account the views of *everyone* in a society. Of course, public opinion surveys and questionnaires fail even more miserably at this than a stock index, while holding on to the illusion of being a democratic measure by claiming to infer overall societal views from a "representative" sample of the entire population. So the question to be answered is whether a stock index can serve as a representative measure of the views of an entire population about the future. Beliefs that an index like the DJIA cannot do this founder on (at least) two fallacies. I call them the *Independence (or Free Will) Fallacy* and the *Homogeneity Fallacy*.

The Independence Fallacy: Most people seem to believe that investors (read: traders and speculators) make their decisions to buy, sell, or hold totally independently of all members of the social group to which they belong. In short, investors exercise total independence of choice in their decisions. If this were true, then the DJIA would indeed only reflect the actions of traders and not the beliefs of an entire society. But of course this "free will" assumption is completely untenable.

The actual order-givers on the financial exchanges do not exist independently of the society of which they are a part. Traders have friends, families, and acquaintances, read newspapers, go to bars, watch television, and, in general, are influenced in the very same ways as other members of society. And it's difficult to imagine that these influences do not affect the decisions traders make about the future direction of financial prices.

So just as a thermometer doesn't have to measure the position, velocity, and interactions of every single air molecule in your living room to produce a very good estimate of the room temperature, a financial market index doesn't have to measure directly the views about the future held by each member of population in order to accurately measure the overall social mood. A representative sample is good enough. And the actions of traders and investors are as good a sample as any, and a lot better than most, as the next fallacy suggests.

The Homogeneity Fallacy: A rather quaint aspect of many public opinion surveys is the tacit belief of pollsters that everyone's view counts equally.

Polls state things like "98% of the population think President Joe Blow is doing a lousy job of managing US foreign policy" or "54% of purchasing managers in Chicago believe the economy will be growing in the next quarter." Implicit in such statements is the assumption that the view of each respondent in the survey counts the same in the measure of how the group feels about President Blow's policies or how the purchasing managers in Chicago feel as a group about the economy. But in the real world we know that opinions, unlike votes, do not count equally. Some people's views are simply more equal than others when it comes to contributing to the overall group sentiment. Some people and some institutions are just more influential in shaping the overall mood of a population than others.

What does this rather obvious statement have to do with use of a financial market index as a sociometer? That's easy. People who are actively engaged in financial matters tend to be people who play a more visible role in the formation of social mood than those who are not so engaged. In other words, what Warren Buffett or an editorial writer at *The New York Times* or even your own stock broker think about the future has a much greater impact on how the society as a whole feels than the view of your barber or your plumber. Some people and some institutions are able to influence a far wider circle of the population with their beliefs than others, and thus weigh-in far more heavily in the formation of the overall social mood. So while Liberty, Equality, and Fraternity are ideals to aspire to, they may not be the dominant forces in the nitty-gritty of everyday life (especially Equality).

Finally, there is the ever-increasing connectedness of the world's financial markets. What happens in New York, for instance, is with high correlation reflected in how stocks open the next day in Asia. And how they close in Asia influences what takes place at the opening bell on Wall Street.

Taking all these factors together, the argument in favor of using a financial market index as a surrogate for the social mood of a population is more compelling than the argument for any of the other candidate sociometers we've considered. So this is the path we'll follow in the remainder of the book. I hasten to emphasize that this does not mean that a financial index is necessarily the best possible sociometer. Nor does it suggest that such an indicator is infallible as a measure of a group's view of the future. It isn't. On the other hand, it seems to work! And since good science like good politics is the art of the possible, we'll invoke that principle until something better comes along.

DOES CULTURE MATTER?

A few years ago as I was just beginning to get into the spirit of the ideas outlined in this book, I was invited to give a series of lectures on complex system theory at the National Supercomputer Laboratory in Petropolis, Brazil, a lovely Germanic-style town in the mountains outside Rio. At the time, I was grappling with the ideas just outlined for using a stock market index as a sociometer to measure the mood of a population. But all the examples I had at hand involved the US stock market. I began wondering whether perhaps the notion of the DJIA as a measure of social mood was something peculiarly American and did not transcend national and cultural boundaries. So I decided to test this possibility using my audience in Brazil.

The basic idea of the test was primitive in the extreme. Before going to Brazil I looked up the monthly averages of the Bovespa Index, the measure of the main Brazilian stock market in Sao Paulo, and plotted it over the period 1992–2006. I then asked my Brazilian friends to tell me what they thought were the most significant social, political, and economic events that took place in Brazil over this period. Peter Kendall at Elliott Wave International was then kind enough to clean up the data and prepare the chart in Figure 2.8, which displays these pivotal events marked on the chart of the Bovespa Index.

The story told by this graphic is compelling. Whenever the social mood in Brazil turned negative, bad things like a currency devaluation, a bank failure, or a presidential impeachment were the *plat du jour*. On the other hand, when the Brazilian mood swung upward and people were looking forward to the future, economic recovery, election of a populist president, and the profitable sale of productive government assets dominated the headlines.

When I visited Petropolis and gave this presentation, it was shortly after the "surprise" election of Lulu DaSilva as Brazil's first socialist president. At the time of that landslide victory in 2002, da Silva had not been taken very seriously as a candidate by the pundits and other talking heads in the Brazilian media. But to one sensitized as to how social mood impacts a collective event like a presidential election, his victory should have been no surprise. As the chart shows, at the time of the election the social mood in Brazil was at a low not seen in more than a decade. Under such circumstances the electorate almost always kicks out the incumbent, just as we saw earlier for US presidential elections. And so it was in Brazil, too.

Figure 2.8 The Bovespa Index and major events in Brazil, 1992–2006.

The Brazilian experiment encouraged my belief in the idea that social mood as a biasing factor in human events was not just a phenomenon confined to American brains. To further support this hypothesis, Figure 2.9 shows a chart for a 50-year period in postwar Japan using the Nikkei

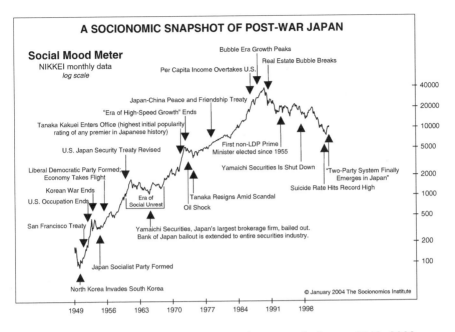

Figure 2.9   The Nikkei Index and major events in Japan, 1949–2000.

index of the Tokyo stock exchange as the sociometer. Again we see that when the Japanese were anticipating a brighter future nice things tended to happen, while pessimism about the future led to just the opposite sorts of events.

As with the election of da Silva in Brazil, a socionomist would have bet heavily on Yukio Hatoyama's victory in the Japanese election at the end of August 2009. As is always the case in times of negative social mood, voter frustration must find an outlet somewhere. In the face of worsening economic conditions in Japan, that frustration involved kicking out the Liberal Democratic Party for only the second time in the post–World War II era, handing a landslide victory to Hatoyama.

A PREEMPTIVE STRIKE ON EVENT CAUSALITY

In a Technology, Entertainment, and Design (TED) Lecture about happiness, Harvard psychologist Daniel Gilbert asks the audience to choose between experiencing two very different events: (1) winning $340 million in the lottery, or (2) having an accident that leaves you a paraplegic. A real no-brainer, right? After posing the alternatives, Gilbert goes on to show data about happiness taken from people who have actually experienced one of these two radically different types of events. He asks the audience to choose which people are happiest a year after either winning the lottery or becoming a paraplegic. Answer: Both groups have essentially the *same* level of happiness with their lives one year after the Big Event!

If happiness, however you define it, in any way reflects how people feel about the future, optimistic or pessimistic, then it's very difficult to reconcile Gilbert's study with any notion that events cause mood. Even in this case of *very* extreme events, whatever momentary mood shift the event brought to the lottery winner or the paraplegic quickly wore off, and the two converged to essentially the same view of their future. Of course, this study is at the level of individuals and you may think that it doesn't apply when we move up to an entire society. Let's see.

The Central Hypothesis of Socionomics discussed in the opening chapter shows an arrow pointing from Social Mood to Social Behaviors/Collective Events, implying a temporal influence moving from how a society feels about the future to the types of events the future is most likely to turn up. But the diagram shows no arrow pointing in the opposite direction from events to mood. In the hundred or so presentations about these ideas I've made over the past few years, I don't think there's been a single time when someone didn't raise this objection. The general thrust is to say something like, "I can believe that the mood of a population somehow impacts events. But surely a dramatic event like 9/11 or a presidential assassination affects how people feel, too. There must be some kind of feedback from events to mood in addition to the arrow you show from mood to events." My guess is that the reader of this book already started formulating just this argument many pages ago. So before we go even one page further into our story, let me lay to rest this natural—but fundamentally flawed—argument once and for all (LOL!) by examining its components. I want to now look much deeper into the extremely contentious claim that *events don't matter*.

In the world of academic finance, and even in the general financial press, one often sees the term *external shock* put forth as a mechanism purporting to explain why a price changes for some good or capital asset like a stock or bond. The belief underlying such statements is that there is a system (the financial market) that sits behind a thick, plate-glass window insulating it from the rest of the world. In some never-quite-described fashion, that outside world broadcasts influences through the glass that serve as external inputs to drive the market from its current price level to a new one. The taken-for-granted background assumption underpinning this phantasmagorically unrealistic picture of the world is that it's possible to decouple the financial markets from the rest of life; they exist literally *outside* the world of everyday human affairs.

As we stated in the book's Overview, this a view of the world taken over almost verbatim from Newtonian physics, classical-mechanics version, in which an observer of a system of moving particles (think planets) can measure the motion of the particles, while remaining *totally* outside the system. Quantum theory stamped paid to this dubious picture of reality, even in physics. And a turn to the human realm shatters this fictitious plate-glass window into a million little pieces. Put simply: There is no outside! Everything—and I mean *everything*—is inside. And it's not even a good approximation to imagine otherwise. So there are no such things as "external" shocks; everything that happens takes place within a system of which the financial markets are but a part, interconnected with everything else.

At first glance, you might think all this interconnectivity undermines the earlier claim that there is no feedback from events to mood. After all, if everything is connected then such a feedback would necessarily have to exist. Let's look just a bit deeper and try to unravel this seeming paradox.

Consider a blockbuster, literally knockdown, event like the 9/11 attack that took out the Twin Towers or a presidential assassination. Chances are you'd say to me, "Surely, John, an event like that would affect your mood." Well, maybe it would affect my immediate *emotional* state (and maybe not). But after years of careful study, Robert Prechter has found that even as a secondary influence the net effect of such events is zero. There is not a shred of evidence (read: data) to suggest that such an event affects the mood of the population as a whole, at all.

Let's illustrate this with an actual example, one of the most dramatic single events of the twentieth century: the assassination of President John F. Kennedy in Dallas on November 22, 1963. As stated earlier, I'll invoke the standing assumption that the social mood is measured by the Dow Jones Industrial Average (DJIA). Figure 2.10 below shows the monthly DJIA over a period that includes November 1963. Now if the social mood (the DJIA) were indeed impacted by that dramatic assassination event, one would expect to see a pronounced dip, or at least change, in the DJIA around that month. Can you pick out November 1963 on this chart? I won't keep

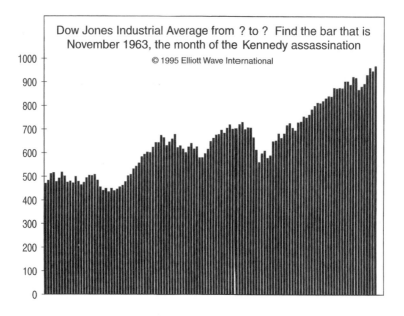

Figure 2.10 The Dow Jones Industrial Average in a period including November 1963.

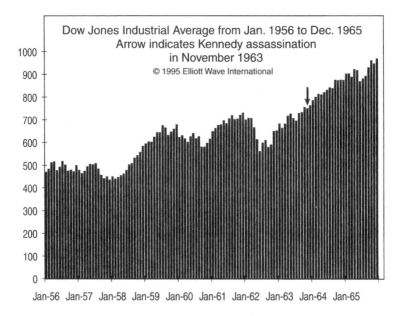

Figure 2.11 The Dow Jones Industrial Average with November 1963 marked with an *arrow*.

you in suspense. Figure 2.11 shows the same chart with November 1963 marked with an arrow. The point here, of course, is that November 1963 is indistinguishable from the months on either side of it. In short, when it comes to affecting social mood, events don't matter—much!

I have shown you only the monthly average and not changes on a shorter timescale. It actually does seem to be the case that there is a kind of "shock factor" at work for such world-shattering events like the Kennedy assassination, and the mood (read: the DJIA) does move in the immediate aftermath of such an event. Presumably, this is a knee-jerk response to the fear and uncertainty that people feel when something out-of-the-blue like the 9/11 attack takes place. But if I had shown you a minute-by-minute chart, what you would have seen is an almost immediate return to "normal" (i.e., the mood prior to the event) within a few minutes or, at the latest, a few hours of the shock.

So the only kind of feedback the data supports is an ultra short-term emotional reaction to the surprise factor of the event. And as the Kennedy assassination illustrates, even for an event with a huge surprise factor like this, the shock-based movement in the social mood very rapidly fades.

Just in case you might be thinking that the Kennedy assassination is an isolated singularity in regard to the impact on social mood of dramatic events, let me cite a few other examples drawn from a report prepared by Mark Jickling of the Congressional Research Service of the US Library of Congress, in response to concerns from Congress that the financial markets might go into free fall upon reopening after 9/11. Jickling published a study on September 14, 2001 that looked at the after-effect on stocks of four dramatic historical events: the attack on Pearl Harbor, the Kennedy assassination, the October 1987 crash, and the Asian financial crisis of 1997. Here are Jickling's findings of the effect of each event on stocks (i.e., the social mood).

Pearl Harbor: Following the Pearl Harbor attack, stock prices declined by about 5% and then recovered slightly. But as the market was in a downtrend at the time anyway, the long-term effect on prices was minimal.

The Kennedy assassination: On the day of the assassination the DJIA fell about 3%. But prices recovered within a week. So the principal effect was restricted to the day of the assassination itself.

The Crash of October 1987: On October 19, 1987 the DJIA fell by a record 23%. The index bounced back over the next 2 days, but remained below the pre-crash level for several months. Jickling then states: "A puzzling aspect of the 1987 episode is the *absence of any concurrent historical event to justify the magnitude of the price drop* [my italics]." This telling remark shows that Jickling himself was caught up in the event causality mindset of how things happen, looking for an external event to account for the shift in the social mood!

The Asian Crisis of 1997: On October 27, 1997 the DJIA fell by a record 554 points, over 7% of its value, but climbed a record 337 points the next day. Again I quote Jickling: "The most plausible explanation for this 'yo-yo' market behavior is that investors in U.S. stocks fell victim to a brief panic. Certainly the Asian crisis could be reasonably construed as a threat to the U.S. economy, but the reaction of the market appeared to be out of proportion to the dimensions of that threat." Here is yet one more indicator of event causality, looking for some outside event as a "cause" of the panic.

And so it goes. Other cases of the same sort include the terrorist attacks on the London subway in 2005 and the Madrid train bombings in 2004. In all such cases, as soon as the event flashed across the news wires the social mood (aka the financial markets) sold off—and then recovered within hours of the event. Put simply, events do not impact the social mood, other than possibly for some very short "panic period" immediately following the public announcement of the event.

Thus far we've laid great emphasis on the fact that there is no feedback from events to mood, and presented a pretty strong case supporting that claim. By now, you might even believe it. But what about other types of feedback from events to the mood of individuals or, perhaps, from social mood back to individual mood? After all, we have already conceded that events certainly do affect the individual's mood in the very short-run, and the collection of the individual moods in turn gives rise (somehow) through the connective structure in a society to the social mood. So perhaps we can sneak-in some event causality via this type of indirect impact on social mood through the medium of individual mood. Or maybe the social mood itself feeds back to individuals, which in turn then leads to a change in social mood. In short, you may argue for a diagram something like the one on the next page.

I think all of us, even me, would accept that one, or even both, of these feedback channels exist. The question before us is whether either of them actually changes the social mood. Let's consider each case in turn:

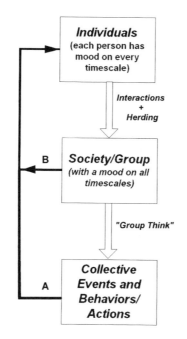

- *Events to Individual Mood (Path A)*: Here the way an individual feels about the future is affected by an event like a terrorist attack. That *might* lead to a change of social mood via the individual's connection to others. But, in fact, we have seen that the occurrence of an event does not change the social mood, at all. Thus, despite the effect the event has on an individual, that effect is not incorporated into the social mood in any measurable, or even observable, way.

- *Social Mood to Individual Mood (Path B)*: Here the conjecture is that a change in social mood, whatever its cause, changes the mood of the individual. Whether that change in individual mood then feeds back to change the social mood depends entirely on how the individual is connected within the group. *If* the existing mood is poised in an unstable state, and *if* the individual's position in the connective structure is such that his or her change in mood coming from the existing mood percolates through the system, then a "mood-changes-mood" pattern might take place. This pattern may indeed be the process by which an existing mood shifts to something different. But the path by which this happens cannot really be known, so the entire feedback process from social mood to individual mood and back again to social mood is an unobservable, unmeasurable pathway in the social system, i.e., it's metaphysical, not at all scientific, and is pretty much equivalent to a random change that happens to set off a cascading process from one social mood to another. But whatever it may be, it certainly is not an *event-caused* change in mood, social or individual.

Since the meltdown of the global financial markets in 2008, economists have scratched their heads wondering how their theories missed this cataclysmic event. One of the consequences of this head-scratching has been a revival of interest in the work of maverick economist Hyman Minsky, who labored in relative obscurity before his death in 1996. Minsky's arguments about the nature of the capitalist economy are eerily prescient and address directly the illusion of event causality. Here's what he had to say about the idea of external shocks and financial collapses.

First of all, Minsky argues that the notion of stability of the economy is itself destabilizing! In other words, the longer a period of economic stability lasts, the more society moves toward being a house of cards built on easy credit and excessive risk. Minsky's basic argument is that in periods of stability investors take on more risk. They eventually borrow too much and overpay for assets (like houses!). Ultimately, speculative borrowers take on so much debt that they cannot even meet the interest payments on their loans without refinancing based on using the increased value of their assets as collateral. But when their collateral begins to decline in value, investors are forced to liquidate thereby precipitating a crash in the economy. In short, the economy ultimately becomes one huge Ponzi scheme whereby old investors get paid off by money "borrowed" from new investors.

Mainstream economics (the event causality type) regards capitalism as essentially stable, moving toward steady growth. In this fictitious world, crises only occur via errors by policymakers (too high interest rates by the Fed, for example), or by external shocks like the 1973 rise in oil prices. Minsky's claim is that the capitalist system itself is subject to crises *from within*, whereby the good times ignite the cycle just described leading to a crash.

There is much more to be said about Minsky's ideas, but for the purposes of this book the above sketch suffices to show how the notion of event causality simply does not exist in the real economy á la Minsky. Everything is *inside* the system; there are no "external shocks" that move the system from good times to a state of crisis. In short, the entire notion of an external shock is an illustion dreamt up by economists to make their models cohere with the framework of Newtonian physics.

The most important lesson emerging from Minsky's work is the central importance of the endogenous nature of social mood. The best place to get a feel for this is working day-to-day in the markets. As we move through a turn, it's a wonder how much worse or better fundamentals get after a swing, up or down, is in place. Remember the move up in the early 1990s? The Savings & Loan crisis was still young, the Persian Gulf War was directly ahead, and the economy was still sinking when stocks turned up. It was the other way around in 2000. So the evidence resides in the turns. They would never happen if a feedback loop from events to mood were in control.

Robert Prechter has given a good account of the infinite regress implicit in this argument of events influencing mood and vice-versa. He says

If events formed a feedback loop with mood, then social trends would never end. Each new extreme in mood in a particular direction would cause more reinforcing actions, and those actions would reinforce that same mood, and so on forever. This is an untenable idea.

LET'S REVIEW THE BIDDING

The material of this chapter is the foundation upon which the rest of the book rests, and the story we've told above has come in many pieces with several twists and turns. Armed with the varied collection of examples presented in the next three chapters, we'll be in a position to see how the Central Hypothesis can be used in action to forecast what is and isn't likely on all timescales. In addition, the outline of a research program required to transform the Central Hypothesis into a formal scientific theory will start to emerge from these deliberations.

Chapter 3

WHY HITS HAPPEN

(Short Timescale Events: A few weeks/months to a few years)

PAINT IT BLACK

I live on a street in the center of Vienna that probably has the highest density of upmarket shops and oh-so-stylish women's fashion emporia of any single street in town. While it would be nice for the sake of everyday infrastructure and convenience to have a hardware store or a decent bookshop mixed in with these fancy boutiques, Persian carpet stores, and very discreet private banks, the display windows of the clothing stores certainly give me an almost daily opportunity to check on one of the most whimsical economic indicators around, the so-called "Skirt Length Index," proposed in the 1960s by Wall Street trader Ralph Rotnem. It goes like this.

If the fashion trend of the day is calling for shorter skirts and dresses, that signifies the economy is booming and the good times roll. But if skirt lengths are heading south, so is the economy. Common sense would argue that when people are optimistic about the future, fashion designers want their creations to reflect that optimism. And one way to do that is to have those elegant ladies with long legs show a bit more of their leggy assets. In the other direction, when we're fearful of the future instead of welcoming it conservatism reigns supreme, and designers reflect this mood by lowering the hemlines. Of course, fashion trends change on an almost daily basis and this kind of economic indicator (if it really is an indicator) is subject to very short-term fluctuations measured by the spring and fall showings in the fashion houses of Milan, Paris, and New York.

J.L. Casti, *Mood Matters*, DOI 10.1007/978-3-642-04835-7_3,

But how does the yo-yo-ing of skirt lengths match up to our measure of social mood, the financial market averages? The tale of this particular tape is shown in Figure 3.1, which displays skirt lengths versus the DJIA over the period 1920–2000, while Figure 3.2a, b shows specific recent examples, the offerings from the Fall 2009 fashion show in New York in the categories black cocktail dresses and over-the-knee-boots, an interesting contemporary "from the other end" variant of the hemline.

The situation didn't change in 2010, either. As Joanne Coles, editor-in-chief of *Marie Claire* declared, "Women don't want dressing up to be so complicated. The hemline dropping is part of that." Designers were also showing more pants than in recent years, when dresses ruled the runway.

These graphics show clearly that the Skirt Length Index is not something to be trifled with. And in trying times like these, one might even encourage fashion designers to contribute to saving the economy by having fashionable ladies wear shorter and shorter skirts. As that well-known fan of short skirts, JFK, put it, "Ask not what your country can do for you, but ask what you can do for your country!" Unfortunately, such a line of argument flies in

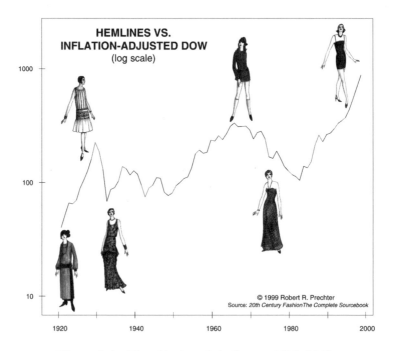

Figure 3.1   The skirt length indicator, 1920–2000.

Figure 3.2   (a) New York fashions Fall 2009, black cocktail dresses. (b) New York fashions Fall 2009, over-the-knee boots.

the face of our Central Hypothesis, since it would suggest that an action (wearing short skirts) can contribute to formation of a view of the future (the social mood). More's the pity, actually. But the world seems to work just the other way around.

A distantly-related, yet still amusing, indicator of when things are hot and when they're not is what New Yorker's call *The Hot Waitress Index*. It works like this. When the mood is positive and birds are singing, extremely attractive young people of both sexes are in great demand for selling everything from exotic cars to premium-brand toothpicks. This leaves the more mundane jobs like waiting on tables to the genetically less well-endowed. But once the mood turns down, all bets are off.

An article in *New York* magazine describes the situation. First the luxury emporia let the boys go, then the less attractive girls. All these hotties then appear out of nowhere. The managers of the restaurants and bars confirm this process saying, "These hot girls that once thrived on the generosity of their friends in the scene for hookups—hosting events, marketing brands, modeling—are now hunting for work."

Of course, to qualify as a leading indicator it must give an indication *in advance* of the economic twists and turns. Interestingly, there is some evidence that the Hot Waitress Index does indeed provide such signals. Here's the line of argument: First of all, employment is traditionally seen as following (lagging) an economic recovery. But "hotness" enjoys a special position in the ranks of the unemployed, one that reverses this direction. As a commodity that's pretty cheap, hotness is likely to be in demand long before demand returns for furloughed airline hostesses or autoworkers. So when you see the homecoming queen serving up shakes at your local Dairy Queen, take heart. The good times will surely soon follow!

But it's not just skirt lengths or hotness that give a hint at what's to come in the economy. Color enters, too. In fact, it's a lot easier to grasp the idea that when the mood is upbeat and people look forward to the future bright, sunny colors are in vogue. However, when the overall outlook on the future takes a turn to the negative, fashionable colors shift to more somber tones with purples, browns, and blacks dominating, as seen in Figure 3.2a, b.

Color and social mood correlate even more clearly in people's choice of color for their car. As the postwar boom gathered steam in the 1950s, the dull shades that dominated car colors in the earlier part of the century began to give way to bolder, brighter colors in step with the optimistic mood of the population about their future. Mark Galasiewski of the Socionomics Institute has done an extensive study of how bright, cheery car colors

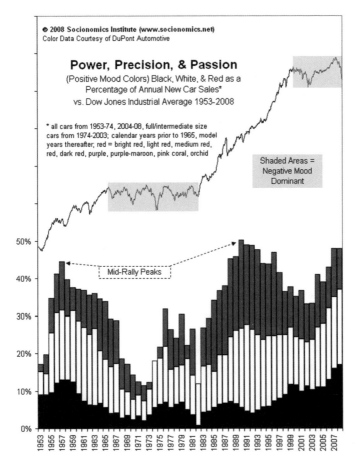

Figure 3.3 Dominant car colors in periods of positive social mood, 1953–2008.

wax and wane with the DJIA. His findings are summarized in Figures 3.3 and 3.4. The first shows that the dominant colors during periods of positive social mood are strong and bold colors like black, white, silver, and red. But in times of negative mood, drab earth tones like green, brown, and beige dominate. In fact, silver was the most popular car color in 2008, just as the social mood was peaking, followed in order by white, black, and red.

It's interesting to compare these color charts with the fashions shown in Figure 3.2a, b. I leave it to the reader to draw his or her own conclusions by streetwise observation of fashionable ladies as to the direction the overall social mood is taking at present.

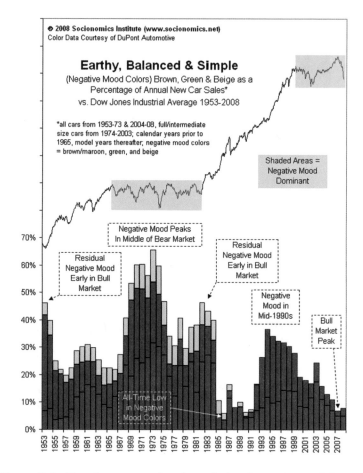

Figure 3.4 Dominant car colors in periods of negative social mood, 1953–2008.

It's but a small step from the runways of the trendy fashion houses and the dwindling number of auto dealers showrooms to the auction rooms of Sotheby's and Christie's. Let's see if something similar might be going on there, too.

Just a few days before the peak of the DJIA in October 2007, the art world was stunned by the sale of a platinum-plated, diamond-encrusted skull created by the British artist Damien Hirst (Figure 3.5). The $100 million paid for this curio was the highest price ever paid for the work of a living artist. And whatever you might think of the symbolism of a cache of diamonds being shaped into an icon of death, you have to admire the sheer audacity of the effort, if not its good taste and judgment. Anyway, the sale of this skull

Figure 3.5   Damien Hirst and the $100 million platinim-plated, diamond-encrusted skull.

encapsulated a mania for the works of living artists, which was one manifestation of the red-hot social mood of the day, and one key extra-market reason why *The Elliott Wave Financial Forecast* called for a huge peak in social mood at that time. But, alas, Hirst's star seems to be falling, as in late 2009 the *London Times* art scene power list showed that Hirst had precipitously dropped from the top spot in the 2008 ratings to number 48.

Unfortunately, though, the negative mood that set in from the time of the skull sale and that continues to this day caught the art world flat-footed. As one art dealer remarked about the huge success of a sale of Hirst's works in September 2008, "At a time when other markets are reeling, the people with the free cash and the will have ignored the storm warnings and the voyage goes on." But the good ship Art seems to have struck a reef, and even though some sales continue to this day to bring extraordinary rewards to a few artists like Hirst, the art houses flogging these pieces are not doing at all well. Figure 3.6 shows this decline for Sotheby's, which provides a preview of what one might expect for the art market as a whole as when the overall social mood goes into free-fall.

If you think this picture is an aberration confined to Sotheby's, consider the fact that worldwide sales of contemporary art at Sotheby's main competitor, Christie's, fell 75% in 2009. All this despite a record $487 million sale of the Yves St. Laurent collection in February 2009. And the crash continues. By October 2009, Bloomberg reported that demand for contemporary art had fallen since the Lehman Brothers bankruptcy in September, with art sales shrinking up to 80% and that in some cases prices for individual pieces falling by more than 50%. Moreover, sales at London's

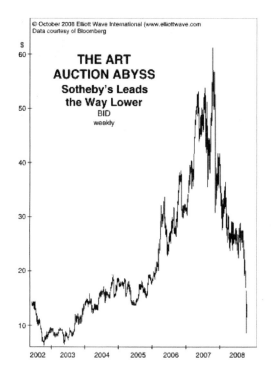

Figure 3.6 Sotheby's stock price, 2002–2008.

contemporary art auction were down 81% from 2008, and three auction houses stopped guaranteeing minimum prices to sellers.

A 1961 Alberto Giacometti sculpture, *Walking Man*, overtook Hirst's *platinum skull* as the most expensive work of art ever sold when it was knocked down for a cool $104 million on February 3, 2010. This action is yet one more sign of a declining mood ahead, as art prices tend to pop to new highs right after big stock peaks. When the door to profitable stock investments closes, wealthy investors look at the bull-market appreciation in art and conclude (erroneously) that a work of art is a more substantive store of value than a stock certificate.

As the social mood blowout of late 2008 gained steam, a niche market of psychologists emerged to tell financial advisors how to handle their customers emotions. As the *Washington Post* stated, investors are "flucutating between fearful, freaked, impulsive, obsessive-compulsive, outraged, disgusted, despondent, contemptuous, and sometimes all of the above." The *Post* article revealed that the methods these behavioral specialists advised are aimed at preventing investors from doing the one thing that would really help alleviate their stress: selling their stocks! The *Post* columnist concludes with the following upbeat remark on the advice pros give investors: "Get them to think about the purpose of saving money, for their dreams and their vision of the future. You don't achieve that dream by selling out at the bottom." The column concluded with this pithy consolation: "If all else fails, we can at least keep dreaming." The *Elliott Wave Financial Forecast* observed in response to this statement, "So there's still no recognition that the dream and the shared passion for it might be the problem, or that clinging tighter ensures that a very real nightmare will get more intense."

But there are other, more foundational, ways of relieving stress than by reading this sort of pop-psychology. Good, old-fashioned sex is one of them, as the following story suggests.

First, let's ask how the social mood affects the type of woman who sets the standard for what is and isn't attractive in a particular era. To address this question, consider the young women gracing the centerfold of that venerable standby of male taste and aspiration, *Playboy* magazine. With tongue-in-cheek, one might call this the "boom-or-bust" theory.

In a "they-got-paid-for-this?" kind of study, in 2002 researchers from Mercyhurst College in Erie, Pennsylvania, determined there is a definite link between the health of the US economy (i.e., the social mood) and the type of woman pictured in the *Playboy* centerfolds. After studying the faces and figures of the Playmates of the Year from 1960 to 2000, the hardworking researchers came to this conclusion: "In boom economic times (positive social mood), men prefer softer, more vulnerable women. When the economy goes bust, those same men prefer stronger-looking gals. In short, we want someone to have fun with when times are good, and we want someone to take care of us—and themselves—when times are bad," said lead researcher Terry F. Pettijohn II in an interview with Reuters. He says his study is the first to consider how social and economic conditions influence changes in body-type preferences. This was done through an annual "Hard Times Index" that tracks changes in US statistics on unemployment, marriage, homicide, and other factors over the four decades considered. Then Professor Pettijohn—and a very lucky undergraduate assistant—precisely measured the key face and body dimensions of the *Playboy* models.

After all this work, the two concluded that when times were tough, the Playmates tended to be older, heavier, and taller with larger waists and bigger waist-to-hip ratios. They also had smaller eyes. When the economy was flush, the models were younger, shorter, and thinner and had smaller waists and bigger eyes. In a family-style book like this, I'll resist the temptation to show a graphic illustrating these points, contenting myself with just stating a few of the more prominent statistical facts emerging from the Pettijohn study.

First of all, there were nine physical features of the Playmates that were measured, including age, bust, waist, hips, height, and body mass index. Among these features, the highest positive correlation with the Hard Times Index were the height, waist-to-hip ratio, and age. This means that these features tended to increase when economic times got hard, as such times are consistent with older "women of substance." In the other direction, when social and economic times got easier, the highest negative correlation was with bust-to-waist ratio and hips. This means that the desirable body type was one that was more slender with a larger bust.

Putting this Playmate example together with the generally declining social mood since 2007, it's an easy call to argue that the next few years will see older, heavier, and larger feminine shapes in the gossip magazines, and that those women will tend to be wearing ankle-length skirts in darker colors.

Styles in clothing and color in cars are but two pieces of a huge mosaic that we might term Fashion. Film and Theatre are two others. What types of films can we expect to see dominate during periods of pessimism? And will people be staying home to watch these performances on their home screens? Or will the lights of Broadway continue to burn brightly with theatre-goers filling the seats as they have over the past couple of decades? Let's have a look.

HOLLYWOOD HITS AND BROADWAY NIGHTS

Awhile back screenwriter William Goldman uttered the classic Hollywood putdown, "Nobody knows anything," referring to all the money, time, and energy studio executives put into trying to build films into blockbusters. And it's not just in the film industry that people don't have a clue as to how hits happen. Goldman's epithet applies equally well to book publishers, fashion houses, car manufacturers, record companies, and virtually every other type of industry whose product's success relies on the fickle tastes of John and Jane Q. Public. To explain this well-known observation, some fall back on the old saw: Timing is everything. Our argument here puts a rather different twist on things: It's not timing, silly, it's the social mood that's everything.

To illustrate the point in more detail, economists Art de Vany and W. David Walls carried out a detailed study of over 2,000 films, concluding that "Revenue forecasts have zero precision." In other words, "Anything can happen," or equivalently, "Nobody knows anything." So while big budgets and megawatt actors may slightly improve a film's chance of success, it's not enough to make the investment worthwhile. De Vany and Walls concluded that the only real determinant of long-run success is plain old word-of-mouth advertising. As they put it, "After a movie opens, the audience decides its fate."

Following in this vein, a story in the *International Herald Tribune* in fall 2009 noted that in the cinema houses today, instead of seeing the audience reflected in the flickering glow from the screen, we see viewer's faces reflected in the glow of their iPhones and Blackberries frantically tapping out text messages to their friends on Twitter and Facebook with a real-time review of the film they're watching and a quick thumbs-up or down as to whether it's worth seeing. Talk about the audience deciding a film's fate!

With the rule of the masses in full force, it's not surprising that the overall social mood can sink a film faster than the *Titanic* (the ship, not the film!) whose theme is out of harmony with the mood prevailing at the time of its release. As Prechter pointed out in 1985, horror films and Disney cartoons are good examples. In 1930–1933 the social mood, as measured by the financial market averages, was about as negative as it's ever been. During that period, five all-time classic horror films were produced: *Frankenstein* and *Dracula* (1931), *Dr. Jekyll and Mr. Hyde* (1932), and *The Mummy* and *King Kong* (1933). People were depressed, the economy had collapsed, and the political storms that were to lead to World War II were brewing. For the next 13 years until the end of the war, films with vampires, Frankenstein monsters, mummies, and the like enjoyed great success. Interestingly, during the short bull market during this period in 1935, the film *Werewolf of London* was introduced—and flopped. But just 6 years later, deep into the bear market of 1941, *The Wolf Man*, which was essentially the same film retitled, soared to the top of the charts.

Residing at just about the opposite end of the psychological spectrum from horror films are the Disney cartoon films that were so popular in the mid-1960s and then again in the 1990s. Figure 3.7 shows how these two types of movies have ebbed and flowed with the overall social mood. We see in the figure the clear relationship between the type of film that resonates

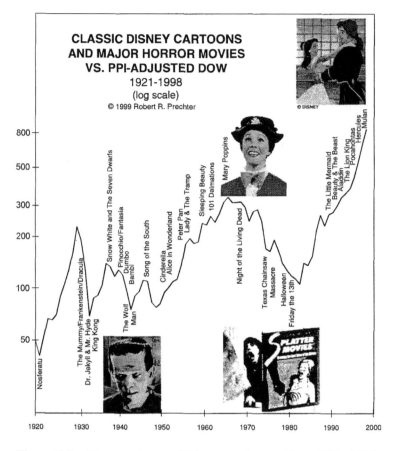

Figure 3.7 The popularity of Disney vs. horror films, 1921–1998.

with the public mood: When the mood is upbeat, so is the film genre that people want to see. In these periods, such as the decades of the 1950s and 1960s, we have films like *Cinderella*, *Sleeping Beauty*, and *Mary Poppins* popping to the top of the charts. On the other hand, periods such as the 1930s and the late 1970s through the mid-1980s see horror films in the ascendancy, giving us such smash hits as *Dracula*, *The Texas Chain Saw Massacre*, and *The Wolf Man*. It's not by accident that a horror writer like Stephen King came into vogue in the mid-1970s, when prior to that period of negative social mood his writings were almost totally ignored. King's breakout novel *Carrie* was published in 1974. Prior to this he had labored in obscurity, even writing under various pen names in the 1960s and early 1970s. And what about today?

About a year after the October 2007 peak in the DJIA, I saw the following headline in the *International Herald Tribune*: "As Boom Turns to Bust, Moviemakers Ponder What Audiences Will Want in Tough Times." The long and the short of the story was that true to its lagging-the-facts tradition, Hollywood was scrambling to produce films that would be "relevant" to a public that was shell-shocked from decimation of their pensions and portfolios, fear of losing their jobs, and general angst over what the future held in store—in other words, a public that was well into a turn from positive to negative social mood.

The number of disaster-type films produced rises and falls in perfect synchrony with the temper of the times. From the 1920s onward, when times are tough and the mood is negative, people want to see disaster; in more tranquil periods, their entertainment dollar goes to musicals and romantic comedies.

During tough times, the films Hollywood puts on offer follow two very separate, but typical, paths: retribution for villains in the executive suites and escapism. For example, in 1987 people saw *Wall Street*, Oliver Stone's portrait of out-of-control traders, which was released just a few weeks after the October crash. On the other side of the fence, some film executives think that during bad times people don't want to see on the screen what they're living through in their daily lives. So, for instance, during the brief mood peak of 1937 in the Great Depression, in addition to horror films people saw films like *The Wizard of Oz* and Frank Capra's *It Happened One Night*, a romantic comedy about a spoiled heiress running away (escaping!) from her family. Both of these films are regularly named on lists of the greatest movies of all time. As Bruce Berman, chairman of Village Roadshow Pictures Entertainment, said recently, "Bad times have spawned some really great movies in the past." But is anybody going to theatres to see these "really great" movies? Let's go back a couple of years to the beginning of the currently unfolding negative social mood and see.

In early July 2005, the film industry reported that the weekend revenues for the top 12 current movies were down 16% from the same weekend in the previous year. This was the 19th straight week that box-office receipts declined from their level the same week a year earlier, surpassing the old record of 17 weeks in a row set in 1985. The report also noted that Hollywood was on course for its lowest ticket sales since the mid-1990s.

It's equally interesting to see the films that top this miserable list: Remakes of escapist and horror fare like *Batman, Star Wars: Episode III,*

Land of the Dead, and yet more remakes of vapid comedies like *Bewitched*, *Love Bug*, and *Herbie: Fully Loaded*. It's as if the film industry was not only alienating potential customers by providing an onerous film-going experience involving the charms of difficult parking, high prices, and an almost endless stream of commercials and coming attractions, but seemed to be flat-out devoid of anything even hinting at a new idea. All the films were either bad remakes of something that worked once before or sequels of the same. Industry faithfuls blamed this decline in revenues at the box office on home videos and DVD sales, citing a rise in DVD sales of 9% since 2004. In short, they argued that people would rather stay home.

As a fascinating aside to this change in film fare, we note the failure by fans some years ago to resurrect the legendary television series *Star Trek*. Trekkies even went to the extent of raising $3 million of private capital toward continuation of the series. But to no avail. Producers decided the market just wasn't there. A lack of public interest in science in general, and most especially in space exploration, is precisely the type of behavior we'd expect to see as the social mood rolls over from positive to negative.

In a similar vein, the last *Star Wars* espisode, *Revenge of the Sith*, is by all accounts the bleakest of the six films in the series. Finally, even the most successful release of the Fourth of July 2005 weekend, Steven Spielberg's remake of *War of the Worlds*, fared poorly. With the most famous director in Hollywood, as well as its highest-paid star, Tom Cruise, this film still grossed only $77.6 million over the long weekend, less than half the amount of the record-holder from the same weekend in 2004, *Spider-Man 2*, which took in over $180 million in its first 6 days of release. It's revealing to observe that Spielberg is on record as stating he would never make a science-fiction film featuring hostile aliens. Evidentally he saw things differently following 9/11, and now talks of the relevance of H. G. Wells' frightening vision of the future for current times. Note that these were films conceived and produced in 2002–2003, a time of strong negative mood, and only appeared on the screen after the mood had turned.

In late 2009, film studios began looking to pay top-tier actors less money as megawatt stars like Denzel Washington, Johnny Depp, Julia Roberts, Russell Crowe, Tom Hanks, and Will Ferrell were all massacred at the box-office in their most recent offerings. "The cratering of films with big stars is astounding," declared Peter Gruber, former chairman of Sony Pictures, who is now a producer and a kind of elder statesman for the industry. As a specific point in case, Will Ferrell bombed in *Land of the Lost*, a $100 million comedy that sold only half that amount in tickets in North America.

It's worth noting that the Christmas weekend in 2009 shattered box-office records with an estimated $278 million in revenues, breaking the previous $253 million record set in July 2009 when *The Dark Knight* was released. This record coincides with the top of the bounce in social mood that began in March 2009. It's unlikely to be topped any time soon, as soon as the social mood turns downward again, most likely with a vengeance in 2010.

In shades of the deepest part of the Great Depression, we see big success for horror films involving teenage vampires. If you'll pardon the bad pun, why are vampires coming back to life now? Nina Dobrev, star of the TV series *The Vampire Diaries*, has the answer: "Vampires don't die. They're always around. They're eternal. And it just seems like right now, people are really responding, and they really like it." As Kevin Depew wisely noted, this quote sounds a lot like the way we might describe credit expansion by the US Federal Reserve from 2008 onward: "Never dying, always around, eternal. And right now people are really responding to it!"

If anything, the situation is getting more severe. Recently, I saw an article in the *Wall Street Journal* titled "Hollywood Destroys the World." It summarizes a raft of post-apocalyptic films due at your local theatre in the coming year or two. Now, though, instead of focusing on staving off the apocalypse, Hollywood is focusing on what happens afterwards! For example, Roland Emmerich, director of the 1996 classic *Independence Day*, which involved a bunch of hostile aliens wiping out human civilization, has now set out to tell us what happened afterwards in his latest production *2012*. In this little gem the Earth will rip apart, fulfilling an ancient Mayan prophecy claiming that the grand evolutionary cycle of Earth will terminate on the winter solstice, December 21, 2012. Again? We'll see!

But if people are staying away from the cinema in droves, perhaps you're thinking they're curling up with a good book instead. Let's again go back to 2005–2006 and look at the record.

A glance at the bestsellers lists at that time was a sobering experience. In popular fiction the lists from both *The New York Times* and the *London Sunday Times* show striking similarities; in fact, there are three books that appear on both lists: *4th of July* by James Patterson and Maxine Paetro, *The Closers* by Michael Connelly, and *A Long Way Down* by Nick Hornby. The first two are police procedural potboilers, while the third is a tale of four strangers who gather together on New Year's Eve to commit suicide. Charming? Entertaining? Depressing? Englightening? I let the reader be the judge as to where these volumes sit on the social mood polarity spectrum.

By way of comparison, here is a sample of the books on the very first *New York Times* bestseller's list in 1935, more than 70 years ago: *Butterfield 8* by John O'Hara, *Studs Lonigan* by James T. Farrell, and *A House Divided* by Pearl S. Buck. Notice a difference?

The situation doesn't get much better when it comes to the nonfiction list. Here the cultural divide between the two sides of the Atlantic is more pronounced. Of the 10 volumes on the British list at the end of 2005, no less than 8 are biographies/celebrity musings-type of books, ranging from a portrait of Elvis by his family (no doubt disappointing its readers by the lack of new information as to his whereabouts) to an exposé of the horrors of that old charmer, Chairman Mao (which sits at the top of the list).

Again, I invite the reader to judge where to place these themes on the positive-to–negative scale laid out in the last chapter. The executive summary, though, is that in the English-speaking world at least, the prevailing social mood as evidenced by the population's taste in books seemed very far from upbeat during the 2002–2003 period when these volumes were contracted.

That's the story for Hollywood and Publisher's Row. How do things stand on Broadway? Is the pattern in popular films present for live performances in the theatre, as well? There seems to be no particular reason to believe that the character and type of plays that will be popular in hard times will differ much from films. But what about actual paying customers? Unlike movies, which "scale" in the sense that you can produce them once and show them millions of times, theatre performances are for the most part one-off events. Moreover, there is a huge secondary market for films in the form of DVDs, a market that's almost totally absent for live theatre performances. Consequently, theatre depends far more heavily on live paying customers in the seats than films. So how does actual attendance at the theatre look?

To cut to the chase, theatre-goers line up at the box office almost in synchrony with the state of their portfolios. This suggests that a downturn in the social mood leads to a corresponding downturn in ticket sales on Broadway, and from late 2000 onward that is precisely what's taken place. It's interesting to note that a study of theatrical offerings at past peaks in social mood show a clear trend to finishing long market advances with flashy but shallow dramas. At the peak of social mood in the late 1990s Stephen Sondheim bemoaned the fact that "You have two kinds of shows on Broadway—revivals and the same kind of musicals over and over again, all spectacles. Broadway today is more than ever about spectacles rather than real drama or real emotions, more about giving audiences an 'experience'." Does that sound like current films or what? The volume *Times Square*, a history of the theatre district in New York, noted that "In the 1930s the frivolity of the 1920s gave way to a serious, if not grim tone." Putting these remarks together, things on Broadway today seem to stand in pretty much the same place on the escapist-serious scale as they stand in Hollywood.

Finally, it's worth noting that another type of social behavior seen during periods of negative mood is gratuitous violence, both on the screen and in the personal behavior of film and music heroes, or perhaps we should say "anti-heroes." Here is just a small sampling of an ever-increasing number of such cases.

Example 1: One of the most talked about films of 2005 was David Cronenberg's thriller *A History of Violence*. This is the story of a seemingly average American family in Nowheresville, USA that is nearly torn apart by brutal violence. As *The New York Times* film critic Manohla Dargis wrote, "The director is taking aim at our [the American] violence-addicted cinema, those self-heroizing, self-rationalizing, self-justifications that U.S. movies sell to the world."

Example 2: In early July 2005, rap diva Kimberly "Lil' Kim" Jones was sentenced to a year and a day in jail for perjury for lying to a grand jury about a shootout that occurred in Manhattan 4 years earlier. This incident by itself is hardly worthy of attention. But it is the tip of an iceberg involving a series of gunfights that have become common among rap groups in the past few years. It also singles out for attention hip-hop radio stations that pump up their ratings by instigating potentially deadly confrontations almong rap groups over the air. The hip-hop press has also made its contribution to this glorification of the "anti-hero" artist by embracing violence and criminal activity as a kind of badge of honor in this corner of the so-called musical world. According to Brent Staples, a member of the editorial board of *The New York Times*, the ethos prevailing in the hip-hop community nowadays glorifies only those who have been shot, committed crimes, or spent time in jail. Only such urban folk heroes it seems are deemed to hold the kind of street credentials making them worthy of studio recording.

In a similar vein, rapper Lil Wayne gave a "farewell" concert for his fans at the end of 2009, just prior to checking in to jail to serve a one-year sentence on a weapons possession conviction.

Example 3: At the beginning of 2006, *Family Media Guide* published a list of the top 10 ultra-violent video games for 2005, listing such lovely attractions for your children as *Resident Evil 4*, a game in which during the first minute of play you can see a woman pinned up on a wall by a pitchfork, *God of War*, in which prisoners are burned alive, and *True Crime: New York City*, in which the player is a NYC policeman who can plant evidence on civilians and shake them down to get extra money.

These examples date from the period when the overall social mood was in its topping phase, so that the negative turn was just starting to develop. Rather closer to the end of the Great Asset Mania we have the 2007 film *The Mist*, which is an adaptation of a Stephen King short story. The story is set in a coastal town in Maine that is covered in a huge mist. The film's main character and his son find themselves stranded in the local supermarket as the mist sets in. Tellingly, as the outside world vanishes in the mist the film's co-star says that "The monsters inside the store turn out to be more frightening than the creatures outside." This sounds eerily like a description of the denizens of the global financial services community today.

The year 2007 also saw the violent and horrific film *Sweeney Todd: The Demon Barber of Fleet Street*. As *The New York Times* film critic stated, the film's creators "systematically subverted" two of the most optimistic genres of popular art: movies and musicals. The review went on to say, "Sweeney is as much a horror film as a musical. It is cruel in its effects and radical in its misanthropy, expressing a breathtakingly, rigorously pessimistic view of human nature"—a perfect description of what one might expect to see at the outset of a rollover in social mood from optimism to pessimism, of which the year 2007 was only the beginning. More generally, this trend toward the dark and horrific in film, theatre, hip-hop, and gangsta rap suggests we can expect to see gangsters, pirates, and other low-life being elevated to the role of folk heroes in the coming years.

Probably the most commonly accessed form of entertainment is music. So let's take a page or two and look at what historical shifts in social mood tell us about the type of music that will be popular and how music sales will fare during tough times.

THE SOUND GOES ON

In the volume *The Wave Principle of Human Social Behavior and the New Science of Socionomics*, Robert Prechter makes the point that all modes of cultural expression tend to move in cycles, but that the particular forms of expression—film, fashion, music, art, literature . . . —that dominate a mode at a given time in a particular culture vary. For instance, jazz music reached a peak of popularity in the late 1930s, while rock music topped in the late 1960s with country-and-western music having its heyday in the 1990s. So while all three types of music fall into the general category "popular music," each form had its moment in the sun to be later shoved off center stage by another form.

Popular music, in general, serves as an excellent illustration of a cultural phenomenon that displays cyclic behavior of just the sort we're interested in here, since music permeates virtually every nook and cranny of daily life. As a result, it's not surprising to find that the themes and lyrical style of the music that's in fashion move synchronously with the overall social mood. Figure 3.8 shows this movement over the period 1950–2009.

From the figure we see that as the social mood moved ever more positive in the 1950s and early 1960s, the dominant lyrical theme was basically "I love everybody," expressed through happy rock-n-roll to surfer songs and through to the so-called British invasion spawned by the Beatles. But when the social mood peaked around 1968–1969, the character of popular music went from upbeat to downbeat almost overnight. This was exemplified by the lyrics in the top song from 1972, *American Pie* by Don MacLean, who sang about ". . . the day the music died." By the end of the 1970s, punk rockers were "singing" a theme of agony and hate for everybody. But from the beginning of the upturn of the social mood in 1982, lyrics became primarily upbeat and happy again.

Prechter's analysis of this correlation between social mood and popular music states that while lyrics are easy to express on the printed page, they are actually the *least* reliable indicator of the the way popular music communicates the social mood. Harmony, melody, the appearance of the performers on-stage, complexity of notes, chord structure, and a lot of other musical and socio-psychological factors surrounding the presentation of pop music are even more reliable signals.

It's an easy forecast to make that the tone of popular music will become increasingly angry over the next few years. Will music sales fall off, too? Let's look back to the period just prior to the breaking of the current crisis to get a hint of what to expect.

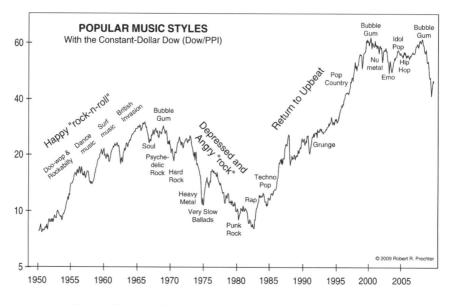

Figure 3.8 Lyrical styles in popular music, 1950–2009.

In early 2006, the Associated Press reported that North American concert attendance declined nearly 12% in the first half of 2005. And this was in the face of the first drop in average ticket price in over a decade, following years of continually increasing prices that turned off many fans. The story notes, however, that major acts like The Rolling Stones, U2, and Paul McCartney still had no problem getting fans to pay considerably over the odds for tickets to their performances. As a spokesman for the concert industry observed, "For the right act, people will pay a lot of money for a good seat." But the "right" acts are fewer and farther between, and the trend for the overall average is heading downward at an accelerating pace. Here's a short tabular summary of the situation in regard to entertainment and the social mood:

Hope	Hubris	Fear	Despair
Family themes	Upbeat	Anti-heroes	Depressing

84

THE SPORTING LIFE

Michael Jordan is certainly one of the most publicized athletes in the history of professional sports. For our purposes, though, the aspect of his career that's of more than passing interest is that it paralleled in uncanny fashion the bull market of the late 1980s and 1990s. In a 2003 study, socionomist Peter Kendall showed that Jordan's first two breaks from the game, an early retirement in November 1993, and his final appearance with the Chicago Bulls in June 1998, occurred right before big drops in stock prices. He reached an agreement to become part owner and an executive in charge with the Washington Wizards on January 14, 2000, the very day the DJIA hit a multi-year high. For Jordan and the Dow, it's been downhill ever since. Both regained some of their old luster in the fall of 2001, when stocks reached a temporary bottom and Jordan announced that he was coming out of retirement. But the social mood of the time, as evidenced in the pattern unfolding in the financial markets, said right off the bat that both efforts would fall short of the old bull-market standards. Jordan's 2-year comeback was an apt metaphor for stocks too, as he displayed flashes of his old brilliance but never regained the form that produced six championships during the 1990s. Of his retirement Jordan said, "It's time. I feel it."

When the bull market ended, the wave patterns of the Dow strongly indicated that Jordan would retire or become less effective. In fact, Kendall predicted this turn of events in print. In another exquisite reflection of the trend in stock prices, Jordan did both, retiring at the front end of stocks' 5-year topping process, returning in time to play less effectively as the first countertrend bounce unfolded, and then retiring again as stocks attempted to reach the old highs and failed. The penultimate act of the drama took place on May 7, 2002, when the Wizards abruptly dumped basketball's most popular figure. "Imagine that, an NBA team showing Michael Jordan the door. Even he was shocked," reported the Associated Press. The DJIA, by the way, topped three trading days later! In stock market terms, Jordan's release was the equivalent of a crash. As it actually turned out, the DJIA ended a rally the following month and then resumed its downward trend. Perhaps it's more than just coincidence that Jordan spent almost his entire career with a team aptly named the "Bulls"! (In March 2010, Jordan purchased the Carolina Bobcats, thus returning to the NBA yet once again. One can only wonder if this will signal yet one more major turn in social mood?)

But Jordan's topsy-turvy career path only mirrors that of the National Basketball Association (NBA) as a whole, as shown in Kendall's diagram

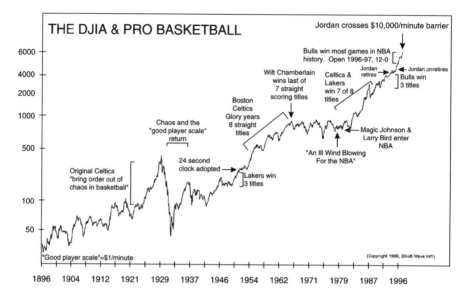

Figure 3.9   History of the National Basketball Association vs. the DJIA.

in Figure 3.9, which traces the ups and downs of the NBA for a 100-year period beginning with the invention of basketball in 1896. The rise of the popularity of the NBA has ebbed and flowed in one-to-one fashion with the long run up in the DJIA over this same period. It's reasonable to expect that the demise of the popularity of professional basketball in particular, and professional sports in general, will continue to follow upon the heels of the bearish downturn in the markets that began in 2007 and that can be expected to continue until the negative mood has run its course. But basketball is not the only sport that will be under siege in a period of resolutely negative social mood.

In a pop culture treatise in 1985, Robert Prechter identified baseball as another bull-market sport. Over the next 20 years many all-time statistical records were set, from home runs to salaries and attendance. In late 2007, however, the DJIA reached its all-time high and *The Elliott Wave Financial Forecast* observed a subtle shift in baseball's popularity.

In mid-December 2007, the Mitchell Report was released giving a full account of the use of performance-enhancing drugs in baseball. As stated by former US senator George Mitchell, lead author of the report, "This has not been an isolated problem. Everyone involved in baseball over the past two decades—commissioners, club officials, the players association and players—share to some extent the responsibility for the steroids era." And on the front page of the December 14, 2007 issue of the newspaper *USA Today*, the headline read "A Collective Failure," the accompanying story amplifying that into a collective experience in "a culture that allowed it." The question is, Why now? Why did this problem, which had been brewing for many years, erupt onto the front page in late 2007 and not several years earlier? A socionomics-based argument gives the answer.

As long as the social mood as measured by the DJIA was rising, baseball fans were ready to ignore rampant steroid use. Commentators argued that use of drugs in baseball is nothing new. As one columnist put it, the steroid scandal will be "vanishing from the public's mind within weeks, maybe days." It's worth noting that during the period of positive social mood, no one stepped-in to try to stop Barry Bonds from breaking the all-time home run record. But on November 15, 2007, Bonds was indicted for lying under oath to a grand jury—*four years earlier!* A *USA Today/Gallup* poll at the time said that 60% of the fans thought baseball should "punish" drug users. What we're seeing here is a sea change in the attitudes of the public toward the widespread cheating that was previously swept under the rug in higher-flying times. Yes, baseball will probably survive. But as the social mood bottoms in the coming years there will almost surely be half empty ballparks and jeremiahs warning of the end of the game. Negative social mood has a way of re-establishing the old order after society becomes so permissive that otherwise solid, law-abiding citizens bend the rules to keep up with the Joneses. One can expect these same self-correcting forces to be at work in the sports arenas around the world over the next few years, as in all other areas of daily life.

To bring some of these scandalous matters up to date, here are a few recent headlines chosen more or less randomly from the *International Herald Tribune's* sports pages that illustrate the basic bias toward anti-heroism in sports during tough times:

April 3, 2008: *Put Bosses in Dock Over Steroid Use*—a column by sportwriter William C. Rhoden calling for shining the steroid spotlight at the baseball team owners as well as the players. Rhoden quotes a number of owners who turned a blind eye to drug use as long as the players put up big numbers that drew fans into the park.

June 12, 2008: *2002 Playoffs Were Rigged, says Disgraced Former Referee*—an article in which former referee Tim Donaghy accuses NBA executives and referees of misconduct and outright manipulation of game results in the 2002 playoffs.

February 7, 2009: *Phelps Gets Suspension and Loses a Sponsor*—an article telling how eight-time Olympic gold medal winning swimmer Michael Phelps was suspended from competition for 3 months and sacked by the cereal-maker Kellogg's as an endorser of their products. In issuing their termination of his contract, Kellogg's spokewoman stated that "Michael's most recent behavior is not consistent with the image of Kellogg," referring to the public photograph of Phelps inhaling from a marijuana pipe.

This list could have been considerably extended with accounts accusing baseball pitcher Roger Clemens, Tour de France contestants, National Football League players, and others of similar misdeeds, including the Tiger Woods circus that began right after Thanksgiving (!) 2009. But why bother? The picture is clear: Negative social mood and cheating in professional sports go together like ham and eggs. The hard-to-escape conclusion is that the sporting world is subject to the very same forces of social mood as other parts of the entertainment world. Here is a summary of how the mood impacts athletes:

Hope	Hubris	Fear	Despair
"Good guy" heroes	Trend intensifies	"Bad guy" anti-heroes	Trend bottoms

The entertainment industry, including professional sports, affects people's everyday lives from the outside, either through attendance at live events or much more frequently through watching events on TV. But social mood also impacts our "inner lives" as we live them out within the walls of our own homes and minds. Relations with our family, mental stress, and even suicide are phenomena that also wax and wane with the overall mood of the population.

MIND–BODY PROBLEMS

By the autumn of 2007, people around the world were experiencing anxiety over being locked into a falling stock market. On September 21, *CNN*

Money announced that "Anxiety was palpable in the financial community all day Monday." And this was prior to the topping out of the DJIA the next month. But the feeling was definitely in the air, as mental health clinics reported substantially increased calls from people seeking help for stress induced by financial pressures. For example, in July 2007 "money worries" accounted for over 75% of the increase in calls to *Hopeline*, a network for people grappling with suicidal impulses. At the same time, the American Psychological Association reported that three-fourth's of Americans were suffering from financial stress.

A sign of the type of aggression, fear, and anger that accompanied the social mood free-fall that began in late 2007 is a store called *Sarah's Smash Shop* in San Diego, where people pay good money to smash up dinner plates, drinking glasses, vases, and even sashimi plates to give vent to their anger. It opened shortly after the peak of the DJIA in October 2007. Sarah's business has been smashing ever since.

In this same vein, we briefly mentioned earlier the emerging niche market for psychologists and behavioral economists who advise investment advisors on how to handle their distraught clientele. Here are a few of the recommendations from these "advisors to advisors" as reported in the *Washington Post*:

- Put the market drop into perspective, maybe mentioning that it is starting to feel like it's bottoming out.

- Listen empathetically, in order to understand the client's irrational behavior and build a bridge to wise action.

- Make the clients see the opportunities that emerge at the bottom of markets and not to miss them.

- Move from the primitive survival mode to the adult, rational-thinking mode [in which selling stocks is always stupid].

And so it goes. One fatuous piece of "advice" after another, all aimed at distracting frazzled investors from doing the one thing that might actually save their sanity: Selling their stocks! (It's amazing that people really pay for this type of hand-holding.)

There's definitely plenty of bad mental health to go around. As the social mood falls off the edge, so does people's grip on sanity it seems. In 1929 when the markets collapsed, one reads stories (almost surely urban myths) of financiers jumping out of their office window and falling to their deaths almost as fast as their portfolios. Thus far, I haven't seen any reports of such dramatic suicides on Wall Street. But this doesn't mean that suicides are not happening, albeit a bit more discreetly. To get some perspective on the matter, let's go back and look at what was happening when the seemingly unrelenting positive social mood was just beginning to run out of gas.

On February 20, 2005, Hunter S. Thompson shot himself at his home in Colorado. Thompson was known as the King of Gonzo Journalism, a style in which the writer is an essential part of the story. But Thompson's story was the story of the mid-1960s, a time when "his" city, Las Vegas, began its ascent with the purchase of the Frontier Hotel in 1967 by Howard Hughes. In terms of social mood, gambling rages in full force during bull markets when the urge to "go for broke" reigns supreme. It continues even in the early phase of a bear-market transition, which accounts for the April 25, 2005 opening of *The Wynn*, the world's most expensive casino.

The symbolic significance of Thompson's self-inflicted death was made even more graphic by a March 9, 2005 headline in *USA Today* proclaiming, "The Suicide Solution Suddenly Seems Trendy." The paper went on to note that the trend is "entirely predictable," although they give no evidence as to *why* it's predictable. Socionomists, though, understand. It is just what one would expect during a turndown in social mood. Whenever markets move down, the public outcry over social issues like suicide skyrockets off the chart, most likely due to not only the increasing number of suicides, but also to a kind of fascination with death in such periods. Witness the issue of euthanasia in the 1970s, the fate of Karen Ann Quinlan in the wake of the 1987 crash, and Jack Kevorkian's notoriety as Dr. Death in 1990, a period when the markets underwent another major downturn. It's no coincidence that the *USA Today* headline appeared just 2 days after a bear-market rally high in the S&P 500.

The situation regarding suicides is not getting any better as the Great Recession gains steam. In September 2009 the European media were filled with stories about a spate of suicides at France Télécom, the country's primary telecommunication service. According to the unions at the company, 23 employees committed suicide since early 2008 in reaction to the stress from a reorganization plan put forward by management. As Labor Minister Xavier Darcos stated, "The first priority is to stop the contagion that's led to this infernal spiral of suicides"—interesting choice of terminology to use the word *contagion*, which speaks pages about the social mood that the minister blames for the suicidal wave at the company. Oddly enough, the suicide rate at France Télécom doesn't differ much from the

French suicide rate as a whole. The national average in 2005 was 17.6 per 100,000 people, which would lead one to expect about 26 suicides in 18 months at a firm the size of France Télécom. This begs the question, of course, as to whether the 2005 rate is the right statistic for 2009. If the national average went down—or moved up—dramatically since 2005, then this comparison is no longer valid. But what is valid is the belief that there is a contagion of mood in the company, and that that localized social mood in the company is at work somehow in these tragic deaths.

In a landmark ruling on January 17, 2006, the US Supreme Court upheld Oregon's assisted-suicide law, stating that the Bush administration had gone beyond its authority in trying to shut down the nation's only program whereby doctors are legally allowed to help people who want to take their own lives. In the court's decision, Justice Anthony Kennedy wrote that the main question the court faced was whether Attorney General John Ashcroft went beyond his powers under the Controlled Substance Act of 1970. As the Justices' opinion stated, "Those limited powers [of the Controlled Substance Act] do not include the ability to declare illegitimate a medical standard of care and treatment of patients that is specifically authorized under state law." I wonder if the Bush administration ever heard of States' Rights (Or human rights either, for that matter)? In every major turn to a negative social mood during the past 80 years, right-to-die legislation has been enacted in one country or another. So the fascination with death in horror films that overcomes a country during bear markets turns up in the right-to-die debate, as well. (Note: This is not just an American phenomenon, as suicide rates are escalating in India as well. In a March 2010 report, India's suicide rate is stated to be one of the highest in the world—and rising, especially among the young and educated.)

Fast-forwarding to the present, it appears ever more apparent that Thompson's suicide put the exclamation point to the period of positive social mood, an exclamation which has engulfed Las Vegas and all its opulent trappings in a downturn that will continue to see gambling palaces like *The Wynn* being shunned like the plague. Just to underscore this point, Figure 3.10 shows the more than 80% decline in the stock price of the *MGM Mirage Hotel* since the beginning of the crisis in October 2007, giving testament to the hotel's name as an indicator of return-on-capital for its investors. And who knows, we may even see Dr. Death or his successor hit the headlines again in the next year or two (perhaps in Oregon?)—a last testament to how social mood biases not only the mind, but also the body.

Figure 3.10 MGM Mirage Hotel stock price, 2003–2008.

An important aspect of lifestyle over the past few decades has been an increasing interest in health foods, "working out," and fitness, in general. What is the socionomic take on this phenomenon? As early as 1985, Robert Prechter surmised that the popularity of exercising moves up and down with the social mood. In the 1920s there was workout audio (on records), but there is no evidence available to suggest that fitness was popular during the Depression or World War II. Instead, cigarettes were popular! In the 1950s and early 1960s comic books had ads featuring body-builders like Charles Atlas. His TV alter ego, Jack LaLaine, built an empire of clubs capitalizing on the growing interest in health and "lookin' good." Not so coincidentally, live wrestling was popular at this time as well. Then in the mid-1960s "working out" became completely uncool. What was *way* cool then was lounging around in love-ins. This lasted through the mid-1970s until release of the film *Rocky* in 1976. Shortly afterwards, jogging became a national craze until about the time its guru, the oddly-named James Fixx, keeled over from a massive coronary in July 1984. In the 1980s fitness centers and TV ads for workout machines became an almost universal mania, and the stock of firms that built and managed such clubs, not to mention

Figure 3.11   Stock price of Bally Corporation, 1998–2003.

those that manufactured Exercycles, Nautilus machines, and the like, soared off into the stratosphere. Figure 3.11 shows the fate of one such firm, the Bally Corporation, in the latter stages of the bull market of the 1990s. It's very worthy of note that Bally is a typical victim of the declining social mood, having filed a Chapter 11 bankruptcy in October 2007, then again in December 2008, after emerging from the earlier bankruptcy. It is indicative of the fact that the ebb and flow of public interest in health and fitness has moved in lock-step with the market—just as socionomics would suggest. When people are optimistic about the future, they want to live long and be in good health. So health and fitness was a big deal in the 1980s and 1990s.

It doesn't take an undiscovered genius to figure out that as the social mood begins to decline, so will an interest in such activities. But in mid-2005, that bastion of what's fit to print, *The New York Times*, stated just the opposite. Quoting a number of fitness club owners and a professor from North Carolina, the article claimed that gym membership increases as times get hard, while good times lead to over-consumption of unhealthy things like aged scotch whisky (ah!) and buttery steak Bordelaise. Of course, the source of this claim is mostly fitness club owners, hardly a group with no interest in seeing you pedaling for dollars (yours to them). Would you consult your stockbroker about the future of the market, someone who has a vested interest in seeing your account remain active and churned as rapidly as possible? I . . . don't . . . think so.

The academic study, on the other hand, comes from a seemingly objective source, Professor Christopher Ruhm of the University of North Carolina at Greensboro. The good professor concludes that health actually improves in hard times, which is believable if your empty wallet forces you to eat less fattening and dangerous foods. He says nothing about whether hard times and good health have anything at all to do with fitness clubs, jogging, or working out, but focuses on normal hazards to health like smoking, diet, and even an increase in auto accidents during good times.

Whither hence working out? Well, perhaps back to boot camp. In early 2004, *The Wall Street Journal* announced that nearly 700 gyms around the country were now offering regimens like military-style workouts and Thai kick-boxing. As Peter Kendall noted, perhaps the future of fitness will be the bearish command, "Drop, You Maggot, and Give Me Twenty."

Family and lifestyle, in general, vis-a-vis the social mood is readily summarized in the following table:

Hope	Hubris	Fear	Despair
Marriage, family	Trend peaks	Divorce, single life	Trend bottoms

Our focus thus far has been on broad swaths of the popular culture domain: fashion, art, entertainment, sports, and personal health, all involving events unfolding on the short timescale of a few weeks to a year or two. But popular culture involves many more things than this, ranging from humor to hobbies to architecture and other fads and fancies. So we close this account of how social mood affects the development of the public taste with an eclectic collection of telegraphic accounts of other aspects of the popular mindset.

POP GOES THE CULTURE

There's hardly an institution in Vienna that's more venerable than the *Kaffeehaus*. And these remnants of monarchy times come in all shapes and sizes, ranging from those frequented by artists and intellectuals to places for businessmen negotiating deals to general-purpose coffee houses where the *hoi polloi* mix with tourists, politicians, actors, and students. In Austria there is a coffeehouse for every 530 people, and Austrians drink 1,000 cups of coffee a year outside their homes and offices. Without a doubt, Vienna is the most focused coffee market in the world. So it was with considerable skepticism that the Viennese viewed the opening of the first Starbucks across the street from the *Staatsoper* in December 2001. Dire predictions that Starbucks had finally overreached itself seem to have been proved wrong, as that "Café Americain" has been a resounding success. Or has it? Let's look a bit deeper into the matter. As always, we go back to the beginning and look at how the Starbucks phenomenon has unfolded in a period of changing social mood.

Starbucks began its expansion in 1987, when it opened outside Seattle. This was just as the social mood underlying the Great Asset Mania was beginning to take off. The firm thrived through the 1990s and up to mid-2006, when its stock began a swoon that continues to this day. The firm's price peaked in November 2006 and has not come close to that level since. As usual, the market is a leading indicator of things to come as sales of Starbuck's "premium coffee experience" are still in free-fall as I write. For example, in early 2009 chief executive and founder Howard Schultz was forced to close 300 stores worldwide and announced a decline in earnings of 69% from the same period a year earlier. So what's going on here?

Well, what's going on is simply that in a period where the dominant mood is fear of the future, coffee hounds in California and South Florida are unwilling to pay top dollar for stimulants, at least the type of stimulants that come in a coffee cup. As Peter Kendall said in 2008, "It's time to wake up and not smell the coffee."

Across the street from the coffee house is the restaurant, another member of an endangered species in periods of negative mood. As Kendall first noted in a study of restaurant performance in 1998, restaurant growth is also a bull-market phenomenon. When people feel optimistic about the future, they like to get out, meet friends, eat well, and in general, have a good time. Restaurants serve as a prime focal point for such activity. But when times get tough, the tape tells a different tale.

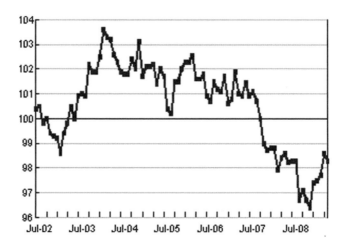

Figure 3.12 Restaurant Performance Index, July 2002–May 2009.

In mid-2008 the National Restaurant Association Performance Index, which tracks the outlook for the entire industry, closed at 98.3. This was the nineteenth consecutive month the Index closed below 100, which is the crossover level between a period of expansion versus a period of contraction for key industry indicators. The chart for the last 6 years is shown in Figure 3.12. Here we see that the level is below its level at the end of the last recession in 2002.

In a July 2009 survey, the percentage of meals eaten out or taken out declined from 50% to 48% from the previous year, a seemingly minor drop. However, 43% of the respondents said they're eating out less, 41% are more price sensitive, and 36% are eating at less pricey places. This trend is not just for family-style restaurants either, as both the Tour d'Argent and Taillevent restaurants in Paris, two of the greatest restaurants in the world, report trying financial times and a dwindling clientele.

One of the unanticipated side benefits to come from the current recession will likely be saying a very long overdue farewell to many of the junkfood emporia lining the strip malls, highways, and byways of the western world.

Another traditional casualty of declining social mood is the character and majesty of public buildings and monuments. Here are a couple of leading examples from New York City, last century's bellwether city for architectural adventurism.

On June 29, 2005 New York City officials unveiled the Freedom Tower, an 82-story structure to be built on the site of the World Trade Center in

Lower Manhattan. With typical understatement Mayor Michael Bloomberg proclaimed, "Construction will climax the greatest comeback in the history of our city." A "comeback"? Maybe. But "great," as in a *great* artistic achievement? The obelisk-shaped tower standing on a 20-story concrete pedestal looks like nothing so much as a gigantic glass paperweight with a toothpick sticking out the top. As critic Nicolai Ouroussoff noted in *The New York Times*, "The darkness at Ground Zero just got a little darker ... somber, oppressive and clumsily conceived, the project is a monument to a society that has turned its back on any notion of cultural openness." In short, the Freedom Tower would be a joke if it were not so laden with symbolic emotional overtones. Unfortunately, one of the things the tower is most symbolic of is how the USA was re-shaping its identity in the post-9/11 era, as an impregnable fortress ready to confront and challenge the entire world outside its borders. All of a sudden, detailed ID checks for visitors, restricted immigration, extension of Homeland Security legislation curtailing individual rights, and all the rest became flavor of the month.

Speaking of architectural shows in New York City in recent years, Ouroussoff remarked in September 2005 that the shows were "safe and predictable, reflecting a growing conservatism in the city's cultural institutions, a byproduct of a society increasingly dominated by corporate values." In such a social and psychological milieu, who would have expected an inspired creation for Ground Zero? Here's another point on the same curve.

Thomas L. Friedman is one of America's most respected journalists and a columnist for *The New York Times*. In early June 2005 he wrote on the same topic of how America is redefining itself in the post-9/11 era. Friedman argues that the country's reaction to the September 11 attacks unintentionally led to an erosion of core elements of American identity. In this article he quotes a British literary agent on her thoughts as she walks past the ugly barriers that now surround the US Embassy in London: "They have these cages all around the embassy now, and these huge concrete blocks, and the whole message is: Go away! That is how people think of America now." Friedman cites conversations with foreign friends about hassles in getting visas to visit the USA nowadays, as he talks about the difficulty American firms have in hiring skilled employees from Pakistan and South Africa, and mentions a conversation with a scientist at MIT who deplores new restrictions imposed by the Pentagon on what nationalities people must have in order to work on their contracts. The take-home message from Friedman's story is that the USA is becoming a state where "the birds don't fly," a state that is isolating itself from the rest of the world on a daily basis.

Does this sound like a shift toward a positive social mood to you? Does it portend broad visions of an open society and architectural monuments to match? These examples certainly do not signal a turn to the positive, at least if the past few years are any indicator.

Care for another example? On the basis of an earlier popularity for outlaws, gangsters, and pirates during the Great Depression, *The Elliott Wave Theorist* forecast in 2003 that pirates would "become popular folk heroes" in the bear market. On August 27, 2005, *The New York Times* reported that following the huge success of the film *Pirates of the Caribbean*, pirate books became runaway hits in local booktores for reasons that defy analysis. The news article asked, "What can explain the allure of pirates? Their role in American history was negligible. As a shaping force in the national culture, they barely exist." Of course, an interest in books and films about pirates isn't quite the same as hijacking a ship in international waters. But it didn't take long before Somali pirates got into the bear market spirit and began hijacking oil tankers, cruise ships, and other targets of opportunity just like in the days of Blackbeard and his band of not-so-merry men. Real-life piracy off the coast of Africa has recently moved these modern folk heroes from the silver screen to the high seas.

The playwright Samuel Beckett once said that "Nothing is funnier than unhappiness." One of the times unhappiness is at its most intense is when a financial bubble explodes in the faces of its creators. When markets are soaring off into the stratosphere, making money and manipulating markets are deadly serious matters. But once the backlash against the bankers and brokers starts, there's nothing funnier than the satirist's pen. So all is not lost during this depressing time, especially if you can maintain a sense of humor. Not even a major economic decline is without its upside. And to acknowledge this truism, the online magazine *Spike.com* produced a list of "The Top Ten Things We're Glad the Recession Killed," from which we extract a few items as a public service in the cause of better mental health:

- *The Five-Dollar Cup of Coffee*—Can we say more than what we already stated earlier about Starbucks? When a cup of java costs more than a gallon of gasoline, you know it's getting pretty late in the day.

- *Celebutant Culture*—Back when credit was flowing like wine and it was cool to live beyond your means, "celebutants" like Paris Hilton and Nicole Richie embodied that spend-till-you-drop lifestyle. But times change. Does your dog really need its own jewelry?

- *Hummers*—More than any other brand of SUVs, Hummer was hardest hit by the economic downturn. Who's going to miss this mass of metallic lard bearing down on them as they enter the freeway?

- *Ridiculous Housing Costs*—There's something out of whack with economic reality when a decrepit two-bedroom shack on the wrong side of the tracks in Miami, LA, or Phoenix carries a million-dollar plus price tag. Get real!

- *"Bling"*—Definition: Bling = wasting money on things that have no useful purpose other than to be an excuse for wasting money. Are you really going to miss precious metal teeth? As Bernard Arnault, chairman of the luxury goods firm LVMH stated recently, "With the crisis, bling bling is passe'."

Amusingly, a headline in *USA Today* in mid-December 2009 asked, "Could the Pop-Culture Mood Mirror Stock Market Changes?" The thrust of this piece was to ask whether trends in popular culture could provide clues as to where stocks were headed. By now, the reader should be hypersensitized to the "backasswards" flow of this line of argument! Like a vampire, though, belief in event causality never seems to die regardless of how many "stakes" of evidence you drive through its heart. But we persevere.

Onward to the next decade.

Chapter 4

WHY WARS, ECONOMIC CYCLES, AND

POLITICAL CRISES HAPPEN

(Medium-timescale Events: A few years to a decade or two)

A TIDE IN THE AFFAIRS OF NATIONS

While aimlessly wandering through the Internet not long ago, I came upon a link to my former employer, The RAND Corporation (named for "*Research and* Development"), which is the prototype for "think tanks" everywhere. Curious as to how RAND has unfolded in the nearly four decades since I worked there, I clicked on the link and was amazed when my screen showed a company headquarters that bore no resemblance to what I recalled from those halcyon days of the mid-1960s. Even a quick glance at these new, hyper-modern buildings made me nostalgic for an earlier era when "deep thoughts for hire" were carried out in far less luxurious surroundings.

When I followed that link, what I expected to see directly across Ocean Avenue from the pier in Santa Monica, California, was a complex of orange-and-cream-colored buildings done in a style that might charitably be termed Mid Twentieth-Century Government Bland. In sharp contrast to the nondescript exterior of those long-gone structures, life inside their walls could only be described as Twenty-First-Century Avant-Garde. Through a lucky turn of events I found myself employed in this heady intellectual environment in the late 1960s as a researcher, trying simultaneously to finance a family and pursue of a doctoral degree in mathematics at the University of Southern California.

RAND's public persona derives principally from its work in military systems analysis, particularly strategic thinking on the use of nuclear weapons, as well as more recent work on a variety of national-security

J.L. Casti, *Mood Matters*, DOI 10.1007/978-3-642-04835-7_4,
© Springer Science+Business Media, LLC 2010

and public-policy issues. However, my own chores were centered on far less controversial, life-threatening, and publicly visible activities, mainly the development of mathematical and computational methods for solving certain classes of differential and integral equations of interest mostly to applied mathematicians, physicists, and engineers. Nevertheless, working in such an environment offered ample opportunity to hear leading thinkers of the day expounding their views at in-house seminars on things like the most cost-effective way to bomb Moscow back into the Stone Age and what the Muscovites might try to do about it.

I recall one RAND seminar at which the speaker advanced a theory regarding the outbreak of warfare. The talk's central claim was to the effect that the transition from harsh words and threats to swords, bombs and bullets is always a chancy affair, almost totally dependent on the capricious whims of a particular national leader. To those of the tide-in-the-affairs-of-men school of historical thought, putting forth a thesis of this kind is tantamount to waving a red flag in front of a charging bull, and the cries of outrage, the posturing, and the ranting and raving (i.e., the discussion) following the seminar were hot and heavy indeed. Since the antipodal claims of chance versus necessity as the driving force behind historical events lie at the heart of whether we can predict and/or explain the outbreak of warfare, it will be of interest for us here if I try to reconstruct a bit of the flavor of the discussion following that RAND seminar.

To protect the innocent and the just plain naïve, let me label the seminar speaker Dr. Chance, reflecting his espousal of the theory that historical events, including the outbreak of war, are driven by the whims of leaders and can be neither predicted nor understood, at least not in the sense that those terms are generally used in the natural sciences. The opposing view, which clearly represented the consensus of the attendees at the seminar, will be presented by Professor Fate. His thesis is that the actual outbreak of war is almost inevitable, provided certain broad sociopolitical, economic, and technological conditions are present. In this view, the actions of individual leaders play a role in initiating war only to the extent of determining the fine details, such as the exact time and place of the initiation of the conflict. But the conflict itself is pretty much foreordained. Let's listen for a moment to a bit of the exchange between our antagonists.

Fate: Now if I understand your position correctly, you are asserting that whether or not a war breaks out is fundamentally just a matter of chance or accidental factors, such as a leader's having a headache or a message's being garbled. Could you give an historical example or two to illustrate your point?

Chance: I'm glad you asked. You've expressed my position exactly, and the history books are full of examples in support of it. For instance, when the bearded King Louis VII of France was married to Eleanor, daughter of a French duke, he received a dowry of two provinces in southern France. Upon coming home from the Crusades, Louis shaved off his beard. Eleanor said he looked ugly without it, and when he refused to let the whiskers grow back again, she divorced him and married King Henry II of England, at which point she demanded her dowry be returned. Understandably, the aggrieved Louis refused, whereupon Henry declared war to regain the provinces by force. This "War of the Whiskers" started in 1152 and raged on for 301 years until peace was finally declared in 1453, following the Battle of Rouen. So here's an example of over three centuries of warfare initiated by the chance shaving-off of a beard. And if you're looking for something closer to home, what about the 1969 half-hour war between Honduras and El Salvador, sparked off by El Salvador's 3–0 victory over the Hondurans in the World Cup soccer playoffs?

F: All very interesting to be sure, but I'm afraid I don't find it very convincing. In fact, Honduras and El Salvador were already at each other's throats over a border dispute. So I think one could convincingly argue that the result of the soccer game was only an excuse, what philosophers would term the "proximate cause" of the war, and that some sort of hostilities were imminent in any case. Surely you must have stronger evidence than this?

C: Allow me to call your attention to one of the major wars of recorded history, the 30 Years' War in Europe, a conflict that lasted from 1618 to 1648 or thereabouts. This war, which resulted in the establishment of the European nation-state system, started on May 23, 1618, when angry Protestant nobles opposing the militantly Catholic King Ferdinand of Bohemia, stormed Ferdinand's castle in Prague and hurled two royal councillors from the window. History records that they survived the 70-foot fall by landing in a pile of horse manure. This chance survival of the councillors in the so-called "Defenestration of Prague" led to the 30 Years' War. You can't get much closer to a chance outbreak of war than this.

F: Surely you're joking. You can't seriously think that if those councillors hadn't been thrown from the window of Hradčany Castle
there would have been no 30 Years' War. What about the turmoil
and strife that Europe was undergoing at the time, things like the
decline of Spain, the political and religious forces pulling apart
the Holy Roman Empire, and the fading of the Habsburgs' dream
of European hegemony? Surely these social, political, and cultural
forces would have led to armed conflict sooner or later, quite independently of a couple of councillors of a minor monarch being
thrown into a pile of horse manure.

C: Of course these sociopolitical and cultural factors enter into play.
I'm only saying that without the chance event, the unpredictable
occurrences of life, things may take an entirely different course—
including warfare *not* breaking out, at all. But now let me turn
the tables and ask *you* to provide some examples to back up your
resolutely deterministic view of the outbreak of war.

F: With pleasure. Let's move into the twentieth century and consider what's probably the most thoroughly studied war of all time:
World War I. An "accident theorist" like you would probably
argue that the war wouldn't have started without the assassination of Archduke Franz Ferdinand in Sarajevo on June 28,
1914. But I think most historians would say that the war was an
almost inevitable outgrowth of several much deeper factors: the
desire of France to recover Alsace-Lorraine, the European system of alliances, Austria's aspiration to dominate the Balkans, the
German Kaiser's aims and ambitions, even the activities of the the
munitions makers and the international bankers. Any one of these
fundamental factors seems more likely to have precipitated World
War I than the archduke's assassination.

C: I see that you include the Kaiser's ambitions on your list of deter-
minants. If this is indeed a significant cause of the war, don't you
think it's more a matter of chance than design that such a kaiser
happened to be sitting on the throne in Germany? If a kaiser with
a less acquisitive nature had been in place, maybe the war would
never have begun.

F: I'm sorry, but I *don't* think another Kaiser would have made one
iota of difference. The actual Kaiser's ambitions were just one of
many such factors, all of which taken together spelled war. The
chance mental makeup of the Kaiser certainly played *some* role.

But it was the overall pattern of politics and life at the time that dragged the world into war, not the personal ambitions of a quirky kaiser. So the *underlying* causes of World War I were inherent in the structure of the international system, even if the immediate cause may indeed have been the archduke's assassination. But I suppose we could go on debating this chance-versus-necessity issue indefinitely. . . .

Indeed they could. So let's try to summarize the diametrically opposed visions of Chance and Fate as to how armed hostilities come jumping out of the closet. To give these competing positions labels, let's call Dr. Chance's arguments the *Cleopatra's Nose Theory* of the outbreak of war, commemorating Mark Antony's well-known infatuation with the temptress of the Nile. As the history books and Hollywood tell it, Antony's chance glandular obsession led him to abandon his naval forces at a crucial juncture in their battle at Actium with the Roman Emperor Octavian in 31 BC, leading to a crushing defeat and centuries of Roman hegemony.

By way of contrast, we can term Professor Fate's deterministic theory of the emergence of war the *Wickedness of Hegel,* reflecting the German philosopher's lifelong antipathy to the idea of explaining human actions in causal terms, thus denying human free will. Here are the main arguments for each position.

- *Cleopatra's Nose (Dr. Chance)*: The events that really shake things up, historically speaking, including those responsible for the outbreak of war, are random, inherently chance occurrences. A typical example is an event like Leon Trotsky's falling victim to a bad cold following a duck-hunting excursion, putting him out of action at a crucial moment in his power struggle with Stalin. According to those admiring the cut of Cleopatra's jib, it's just this sort of caprice that turns the tide of events and accounts for revolutions, economic collapses, and all the other discontinuties of history—including the outbreak of war.

- *Wickedness of Hegel (Prof. Fate)*: Big events don't just "happen." Historical discontinuities like the outbreak of war have deep causal patterns, not random ones, and could not have happened differently unless something major in the causes had also been different. In this view, the nightmare quality of events of the sort found in Kafka's novels lies in the fact that nothing that happens has any apparent cause, or at least any cause that can be ascertained. Thus, in the context of warfare events like Trotsky's illness or Cleopatra's charms are just accidents of history and do not enter into any rational analysis of the onset of fighting. In short, there are genuine laws of history involving broadly based properties of states like their economic, social, military, and political strength, and it's to these laws that we must look in attempting to understand and predict when and where conflict will break out.

It's evident, I think, that our antagonists Chance and Fate would argue for radically different positions on the matters of predicting and/or explaining how wars come about. Chance would undoubtedly say that predicting a war is flatly impossible, since to do so would involve predicting the occurrence of what is by definition an unpredictable event. He would further claim that such chance occurrences, should they happen at a time and place when competing forces are delicately balanced, can tilt the scales of history and spark off a chain of events that may lead to a totally different world. Thus, in Chance's world there is neither prediction nor explanation of war in any meaningful sense of those words.

On the other side of the seminar room, Fate claims that prediction is in principle possible, at least in a statistical sense. Furthermore, he makes the stronger claim that the explanations or causes of war are very plain for all to see, residing in sociopolitical and military factors that have remained unchanged for millennia. Thus, in Fate's deterministic view of the world the only obstacles to the accurate prediction and explanation of wars are practical, not theoretical, clustering about the twin barriers of insufficient data and inadequate understanding of human behavioral patterns. In the opinion of those sharing Fate's convictions, these barriers will gradually melt away with time, ultimately allowing us to anticipate the outbreak of warfare as one of the fruits of a better understanding of human nature.

Clearly, both Chance and Fate are caricatures. I have drawn their respective positions in broad strokes, using only the brightest colors for the sake of making evident the dichotomy between the chance and necessity hypotheses regarding armed conflict. It should be evident to the reader by now that my position on this issue of warfare falls unambiguously on the side of Fate, not Chance.

Before taking a look at some evidence supporting this view, let me note that wars come in many shapes, sizes, and time frames. For the sake of our story, the outbreak of war will appear as a topic in the next chapter as well, reflecting the fact that some wars simply take longer to unfold than others. Note here the term *unfold*, since it is the unfolding time between the development of a shift in social mood and the outbreak of war that determines the time frame, not the length of the war itself. With this point in mind, let's turn now to the battlefield.

In 1940 Ed Dewey started the Foundation for the Study of Cycles in Pittsburgh, Pennsylvania. Over the next several decades he studied a bewildering array of cyclic phenomena, discovering everything from an 8-year cycle in cigarette production to a 33-month cycle in the issuance of residential building construction permits. After many years of cycles research, Dewey reported that most—but not all—cycles had periods that were related to each other by small integer multiples that are powers or products of 2 and 3. So, for instance, the 8-year (= 96-month) cigarette production cycle is nearly three times longer than the 33-month building permit cycle. Of particular interest to us here is the connection Dewey established between earlier work by Raymond Wheeler relating climate to war and the cyclic pattern of international conflicts over many *centuries*. Here is a brief look at a few of these results.

Initially, Dewey constructed a kind of timeline of international battles by counting for each year the number of small, medium, and large battles that took place. He then drew a chart for a given year by placing a short block on top of that year for small engagements, a medium-sized block for moderately large battles, and a long block for major conflicts. These blocks were then placed on top of each other for the year in question. Carrying out this operation for many years, Dewey constructed a chart that looks a lot like the skyline of an urban metropolis measuring the ups and downs of warfare through the ages. By 1952, Dewey had isolated four cycles in international warfare of lengths 142, 57, $22\frac{1}{2}$, and $11\frac{1}{5}$ years. For example, the 142-year cycle suggests that there should be a greater than average number of battles from 1914 to 1985, with a corresponding lower than average number from 1985 to 2056. At the other end of the scale is Dewey's $11\frac{1}{5}$-year cycle, which he traced back to 600 BC. He estimates that a pattern showing this sort of cyclicity could have arisen by chance with a likelihood of only one time in 500.

By combining all four war cycles, Dewey predicted in 1952 that there would be tough times in the 1960s, with a relatively peaceful period in the mid-1970s. As it actually happened, there were small skirmishes in the early sixties between The Netherlands and Indonesia, Syria and Egypt, and Tibet and China. But the Vietnam War in the late 1960s completely overshadowed these relatively minor tiffs. And, as we now know, the mid-1970s were indeed a relatively peaceful period following the end of the Vietnam War in 1975.

In his discussion of these cycles of war and peace, Dewey noted that international wars are sensitive to both biological and economic cycles. So war is both a biological and an economic event. It's extremely interesting for our purposes here to quote Dewey's final comments on the theme of wars and how they arise:

> In my own mind I picture the space in which we live as filled with forces that alternately stimulate and depress all human beings—make them more or less optimistic, or make them more or less fearful. These forces do not *control* us, they merely *influence* us. They create a *climate* that is sometimes more favorable to war and sometimes less favorable. War will come without the stimulus of these forces and wars will be avoided in spite of these stimuli, but, *on the average*, the probabilities of war are greater when the "climate" is right.

Dewey's statement is about as good a description of how social mood biases events as any I could possibly imagine. So wars, just like climatic phenomena, follow their own cycles and appear in a regular pattern that is predictable in just the same probabilistic sense that we can predict droughts, ice ages, and other aspects of the climate. Wars and climate are global phenomena that Dewey and others have shown move in statistically predictable ways. But this also applies to far less grand phenomena arising in the everyday cultural arena, as we saw in the last chapter when we looked at fashions in clothing, film, and music. So it's not just fashions that have trends; it's *all* human social activity.

The work by Dewey shows there are cycles in the military affairs of nations. No doubt about that. And his testament certainly speaks of "forces" that affect human beings, disposing them toward or away from war. Our view here is that these forces are not as mysterious as Dewey suggests, but reside within the collective human psyche, what we're calling the social mood. If that is indeed the case, there should be some evidence on record of social mood biasing the tendency of nations toward belligerent or peaceful behavior. As Dewey's work suggested some time ago, it's reasonable to wonder how things stand nowadays. Does a dark mood dispose nations and factions to dark deeds? Or ??

While events cannot be used to forecast the market, they can be used to help identify the position of social mood, particularly when they reach extremes. The events of September 11, 2001, are a good example. On that day, *The Elliott Wave Theorist* used the extreme nature of the violence and the anxiety that it produced, in conjunction with a culminating wave pattern in the stock market, to call for the S&P 500 to make one more new low and then turn up in a "substantial rally." This position was an extremely contrary view at that time, as the New York Stock Exchange was shut down. When it stayed that way for several days, many insisted that it would plunge further upon reopening. But five days after it reopened stocks did, in fact, bottom and go on to the biggest rally since 2000. The larger message behind the terror, though, is that it marked a return to the negative mood manifestations at the prior bear market of similar magnitude, which ran from 1966 to 1982. As Robert Prechter put it at the time,

Aside from a slew of terrorist incidents in the 1910s, a decade domi-nated by a long bear market in stocks, the idea of "terrorism" as a social force began with the 1966-1982 bear market. A 10-volume chronology of American social history over the last 100 years *(American Decades,* Manly Inc.) lists no acts of terrorism in the 1920s, 1930s, 1940s, 1950s or the 1960s. In early 1970, when the Dow was in year four and the Value Line Composite in year two of a bear market, the age of terror-ism began. An Arab terrorist strike killed one person and injured 23 in Munich. On February 21, 1970, another group of suspected Arabian ter-rorists hijacked a Swissair flight and crashed it. On September 6-9, 1970, Palestinian terrorists hijacked five planes and blew up three. Unlike the earlier period of "scattered terrorist bombings" in the 1910s, this era was distinguished by the frequency, extent and organization behind the attacks. As the bear market continued, the events grew in size and com-plexity. The increasing sophistication of terrorist tactics and their rising political power is revealed by the progression from the scattered strikes in 1970 to the PLO raid on the Olympic Village in Munich in 1972 and eventually to a state-sponsored takeover of the U.S. embassy in Iran in 1979. The worst of it ended approximately when the bear market did, as Iran finally released U.S. hostages in January 1981. In those first terror-ist acts committed against the United States, we caught a glimpse of the style of conflicts that we would have to endure when the bear market of next bigger size arrived, which it certainly has.

Prechter noted as early as 1995 that "Foreigners will commit terrorist acts on U.S. soil." Noting that the events of 9/11 were "more dramatic and on much bigger scale" than those that occurred during the bear market of the 1970s was confirmation that the bear market of the 2000s was also of a higher degree and would ultimately resume, which it did. On that fateful day, dramatic loss demonstrated that "the world that we know has changed." This was a correct assessment. But few saw it within the full context to which Prechter directed his readers when he made a critical socionomic distinction: The change did not come on 9/11. The change came "between January 14 and March 24, 2000," a two-month period during which the three major US stock indexes signaled the end of a 216-year uptrend.

As Prechter notes in his books and our discussion in the next chapter will further elaborate, wars, the ultimate expression of collective human conflict, generally occur at or even after final bottoms. So it fits that a few weeks after such a bottom in March 2003, the United States attacked Iraq. Hostilities then generally continue after the mood turns positive, still working off that earlier negative extreme. The Civil War and World War II are classic examples as they were conducted after major lows in 1857 and 1942.

THE BUSINESS OF BUSINESS

Common wisdom holds that wars are good for business, as they free up capital for the manufacture of a broad spectrum of "toys for boys," such as tanks, fighter planes, explosives, grenades, and high-tech missiles. We've now seen that wars themselves break out in a quite regular, statistically predictable pattern. What about business? If it follows war, business too should show much the same cycles.

In his memoirs, President Herbert Hoover spoke about the consensus of economists prior to the Great Depression, noting about the collective wisdom of these luminaries that "With growing optimism, they gave birth to a foolish idea called the 'New Economic Era'. That notion spread over the whole country. We were assured we were in a new period where the old laws of economics no longer applied." On July 30, 1998, seven decades after Hoover's economists had announced their suspension of the iron laws of economics, a noted MIT economist wrote in the *Wall Street Journal,* "This expansion will last forever," referring to the "New Economy" boom of the 1990s. Most interestingly, this opinion was the *consensus,* not an isolated fantasy, as 98% of economists polled at the end of 2000 said they expected continued expansion throughout the coming year. Now if either Hoover's or the *Wall Street Journal*'s "experts" had the slightest idea what they were talking about, they could never have made such amazing claims for the simple reason that economies run in cycles just like every other aspect of natural and human life. And if you don't believe this, let me give you just a short account of the actual observed facts, not fanciful personal opinions.

Put simply, business cycles are recurring fluctuations in business activity, generally measured by the ups and downs of *real income,* defined as gross domestic product (GDP) adjusted for inflation. One such business cycle is defined to be a period of economic contraction followed by a period of economic expansion. The general picture is seen in Figure 4.1. The reason for such cycles is that the rate of economic growth temporarily exceeds the growth in the ability of the economy to produce goods and services. Eventually the cost of borrowing to finance this expansion rises, followed by a corresponding rise in the price of resources. Profit margins for businesses then decline, leading to a slowdown in business activity.

There are several different types of business cycles, ranging from the Kitchen inventory cycle of 3–5 years to the very long Kondratiev cycle of 45–60 years that we will discuss in the next chapter. But it is the 3–5-year cycle that people generally mean when they refer to the "business" cycle.

Some schools of economic thought, especially those associated with the so-called Austrian School of economics, argue that the business cycle is not an intrinsic feature of a free-market economy but is caused by government manipulation of the money supply through interest rate maneuvers orchestrated by the central banks. In this view, the "dot.com" bubble and its subsequent collapse were caused by artificially low interest rates that led to an unsustainable increase of credit that finally had to be squeezed out of the economic system. Now *that* sounds familiar! It's of interest to

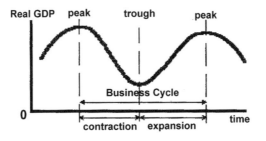

Figure 4.1 A typical business cycle.

note that the 2004 Nobel Memorial Prize in Economics was awarded to Finn E. Kydland and Edward C. Prescott for their recognition that it is fluctuations in supply, in other words, technology shocks, that create these cycles.

Interestingly, neoclassical economics speaks mostly about economic equilbria, where supply and demand curves intersect. Theoretically then, the economy should "equilibrate" and simply stop dead in its tracks at the moment supply and demand are in balance. Of course, this never happens—even approximately. Instead we see patterns that don't strictly repeat but at least show a continuing family resemblance to each other. And that family is a curve with a definite cyclical structure. So where do these cycles come from? One source is "friction," i.e., time lags in the process of transmitting information about demands, prices, labor supply, and all the rest from one part of an economy to another. Let's have a quick look at how this informational friction enters into the creation of cycles.

Assume that initially the price of some commodity, say the now unlamented sport utility vehicles (SUVs), stood in equilibrium and then began to rise due to excess consumer demand (luckily for this example, the argument also works if the demand decreases by just changing the sign of everything from plus to minus and vice versa). While the excess demand does indeed work to increase the price, this process *takes time* since SUVs, like skyscrapers, cannot be built overnight. Thus, the price of SUVs will keep on rising until the excess demand totally disappears. At this point, the demand and supply are in equilibrium. But the price has continuously increased and so the current price cannot be in equilibrium because it is now higher than its original level.

Now the reverse process sets in. The high price of SUVs continues to stimulate supply but restricts demand, which is the root cause of the emergence of an excessive supply of SUVs. This oversupply then acts to exert downward pressure on the price via mechanisms like dealer discounts, rebates, and so forth, a process that itself takes time until the price returns to its original level. At that point, the excess demand has been replaced by an excess supply, which in turn further decreases the price of SUVs, reducing excess supply and leading to a lower price for SUVs than the original level. And so it goes, the price and demand/supply dancing in a cyclic fashion around the theoretical equilibrium SUV price and demand levels. The entire process is shown in Figure 4.2.

In this simple economic situation, it's clear that what causes the oscillatory behavior is the time lag between when the excess demand of SUVs

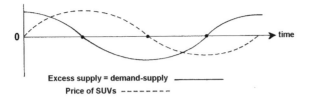

Figure 4.2 Price of SUVs and excess supply oscillating with time lags.

arises and when the effect of the increased price change can actually exert influence to reduce that demand. *If* SUVs could be built instantaneously, that is, if the time lag were zero and price changes could be immediately transmitted through the economic system to affect the supply and demand of these vehicles, then indeed price and demand/supply would remain perpetually in Adam Smith's state of theoretical equilibrium. But nature hates infinities and infinitesimals alike, and here the time lag is not even approximately zero. Consequently, cyclic behavior is hardwired into the economic system. And this goes well beyond economics.

Time lags of the sort outlined here are part and parcel of every real-world dynamical process. Ecologists are equally familiar with such behavior in simple systems involving predators feeding on prey, where we find the world of Adam Smith dressed up to look like the world of sharks and minnows (or swimmers) instead of SUVs and dollars. In all cases, wherever there's competition for a scarce but replenishable resource, be it food, money, sex, fame, or power, the pattern of nature and humans is cyclical, not static. If you want a one-word explanation for why virtually all dynamical processes display wave-like behavior instead of just standing pat, the reason is ... time lags, or what amounts to the same thing, friction. The path from cause to effect takes time. Information doesn't instantly get assimilated into the price of a stock; hostilities don't erupt immediately upon the decision of the high command to initiate a war; public taste in films doesn't emerge on demand with the first showing of a new film genre, etc.

So it is the time lag from the first cause of an event in a chain of connection to the the event's actual occurrence that is the principal generator of cycles. A common example of this process shows up when you speak over a bad phone connection. The delay between when you speak and when the party at the other end responds very often results in a comical back-and-forth, out of synchrony conversation in which each party is making

statements like "Can you hear me?" and "What did you say?" just prior
to receiving the message that answers the query. A similar kind of out-of-
sync cycle takes place with a computer keyboard whose response rate on
the screen is too slow for human comfort. You press the key and expect a
more-or-less immediate reaction. But if the appearance of the letter on the
screen is delayed too long, you press the next key and the next until there
is a big backlog in the buffer. This backlog makes it very difficult to avoid
a kind of destructive cycle that ends up corrupting the entire line of thought
(and screen text!) you're trying to express.

But this excursion into Economics 101 is drifting off our main theme of
investigating how the social mood influences the affairs of business-oriented
folks and their corporations.

In Chapter 2 we briefly discussed the work of maverick economist
Hyman Minsky about the origin and disappearance of financial crises.
Specifically, we presented Minsky's thesis that stability in the economy is
by itself destabilizing. As a result, the capitalist system carries within itself
the seeds of its own destruction—and resurrection. As Minsky himself put
it in 1974: "A fundamental characteristic of our economy is that the financial system swings between robustness and fragility and these swings are an
integral part of the process that generates business cycles." Minsky stated it
another way when he said, "There is nothing wrong with macroeconomics
that another depression won't cure."

The work by Minsky relates directly to the matter of time lags and how
they give rise to cycles via the debt phases we described earlier. The passage
from one phase to another necessarily requires time, generally many months
to a few years, and those time lags between when the loose/tight credit
comes available and its effect moves in or out of the asset market. This
creates friction in the system that ultimately forces a crash or a revival.
Now for a look at the role of social mood in these matters.

One way the mood shows up in particularly stark form is in legal challenges by governments to the way business is carried out. These lawsuits
revolve around issues of antitrust violations, as well as fraudulent actions
that bilk investors out of their hard-earned cash. As the reader will surely
suspect by now, such suits tend to appear when an unrelenting positive
mood starts running out of gas and vanish like a trickle of water in the
desert when people begin looking forward to the future again. Here are a
few juicy examples of this sad—but predictable—story.

The US Justice Department sued Microsoft, the world's most successful
company, on May 18, 1998, one month after the New York Stock Exchange

index had reached a major peak. Of special note is that the index had risen steadily for 24 straight years! Then on April 3, 2000, a bare one week following the high in the NASDAQ 100 index, courts agreed with a government ruling that Microsoft had indeed willingly violated the antitrust laws. The judge in the case, Thomas Penfield Jackson, spoke with an almost evangelical fervor doing everything within his power to break up the most successful company of all time (as measured by percentage gain in value per year). The judge, of course, was simply representing "the people," and his anti-Microsoft passion is nothing more than a reflection of the transiting social mood of the populace at the time. But this was nothing new.

The Sherman Act, which outlaws "trusts," was passed by the US Congress in 1890, a scant one year after a new peak in stock prices. This pattern persists throughout the entire twentieth century. Front-page legal suits consistently come immediately upon the heels of market peaks (i.e., peaks in social mood). Settlements, on the other hand, regularly occur just after market troughs (when a negative mood has bottomed-out and a positive one is just beginning). The 1969 IBM suit is a good case in point.

In January 1969, one month after the biggest bullmarket top since the crash of 1929, the Justice Department sued IBM on more or less the same grounds as the suit 19 years later against Microsoft. Then, in a dramatic about-face, in 1984, at the beginning of the longest bull market in history, the government settled the suit by the simple expedient of dropping the case!

If you were of a conventional frame of mind, you might think the government's filing of such a suit would seriously damage a company's value as measured by its stock price. If you think that, think again! Microsoft's stock price more than *tripled* in the weeks following filing of the suit. Even more interesting, in a November 1999 "finding of fact" by Judge Jackson, who in his customary bombastic style vilified Microsoft stating they are "... a monopoly with practices that wounded competition and consumers and hurt innovation," the market carefully evaluated the judge's words and decided to more than *double* the share price. So much for Judge Jackson!

To bring antitrust matters a bit more up to date, think about the resignation of Eric Schmidt, CEO of Google, from the board of directors of Apple. Schmidt stepped down on August 3, 2009 as the US Federal Trade Commission was investigating whether Google and Apple were breaking antitrust laws by sharing board members. According to antitrust lawyer Samuel Miller, who represented the government in the aforementioned

Microsoft case, Google's Android phone software and Chrome operating system were becoming direct competitors to Apple's products, increasing scrutiny of the firms. Unnamed sources stated that the Justice Department was also looking at collusion in hiring practices at other Silicon Valley firms.

In 2000 Robert Prechter characterized the overall phenomenon hitting Google today when he stated, "During a bull market, the social mood is directed toward rewarding achievement; during a bear market it is directed at punishing it." When one looks at Google's dominance of the Internet search market, along with its 779% growth from 2004 to 2007, it's no surprise to see it sitting directly in the crosshairs of the Justice Department in these times of rapidly declining social mood.

Let me close this brief discussion of antitrust actions with a mini "pop quiz" for the reader. Why does the government bring antitrust suits after the stock market has risen for years or even decades to a level of massive overvaluation rather than at any other time? Those now steeped in socionomic thinking will have no trouble understanding what's going on here, and will certainly be able to easily unravel this seeming conundrum.

Now we turn to a closely related matter, one dear to the hearts of just about every investor nowadays in the aftermath of the biggest financial crisis since the 1930s. But this time we'll walk the legislative side of the street rather than the judicial, looking at the way legislators continually shift their stance on how to best regulate the financial services industry in order to "protect" the small investor.

In their typical self-serving fashion, legislators pander to the mood of the public in almost every action they take. There's no area in which that is more evident than in the so-called "protective" legislation enacted to salve the wounds of investors wiped out by the flow of the market and/or by rapacious investment "advisors." Given the predilection of "mom-and-pop" investors to believe that those who succeed in the market when they themselves fail must be doing it by chicanery and corrupt business practices, legislators necessarily have to acknowledge these sentiments by passing punitive laws whose surface appearance is to protect small investors from such predators. On the other hand, when even the corner shoeshine boy is serving up hot tips on the markets no one wants to be restricted in what they can do. In such bullmarket periods, legislators, sensitive as always to where the votes are, repeal and/or rewrite previous legislation to free the hands of the investment community.

Tracing the ebb and flow of such legislation, economists John Nofsinger and Kenneth Kim, expanding on earlier work of Peter Kendall and Robert Prechter, identified a number of important laws designed to protect investors. The dates of these laws, their purpose, and most interestingly what was going on in the world of finance shortly before the law was passed are shown in Figure 4.3. What immediately strikes the eye in this chart is that after each period of scandals and negative social mood, Congress immediately stepped-in to take visible action to punish in some way the perceived transgressors. In fact, investor protection legislation was enacted after every one of the major market declines in the twentieth century! As any decent socionomist would expect, these very same laws get weakened, if not repealed all together, when the social winds start blowing in the opposite direction. Figure 4.4 gives the story.

Act	Purpose	Preceded By
1933 and 1934 Banking Act and Securities Exchange Act	Separates commercial and investment banking, creates SEC as market regulators	Stock market crash of 1929 and ensuing bear market removes nearly 90% of Dow value
1940 Investment Company Act and Investment Advisors Act	Regulates investment companies and advisors	Market decline of 25% from October 1939 to May 1940
1970 Securities Investor Protection Act	Creates Securities Investor Protection Corporation and insurance from broker defaults	Market decline of 30% from April 1969 to June 1970
1974 Employee Retirement Income Security Act	Regulates pension funds	Long bear market from December 1972 to September 1974 takes the Dow down 40%
1988 Insider Trading and Securities Fraud Enforcement Act	Increases penalties and liabilities for insider trading and fraudulent activities	Stock market crash of 1987 takes Dow down over 40%
2002 Public Company Accounting Reform and Investor Protection Act	Increases regulation of auditors, lengthens punishment for white collar crimes, and creates more corporate fraud laws	2½ year bear market reduced Dow by 35%, Nasdaq declines 75%

Figure 4.3   Legislation passed to protect investors.

Action	Purpose	Preceded By
1927 Government agency policy allowing commercial banks to issue securities	Allows commercial banks into investment banking activities	Stock market rose over 200% from 1925 to 1928
1995 Private Securities Litigation Reform Act	Limits the ability and available damages of investors suing for corporate fraud	Dow increased 60% between 1993 and 1995
1998 Securities Litigation Uniform Standards Act	Precludes plaintiffs from bringing securities actions in state courts	Dow increased 125% from 1996 to 1999
1999 Financial Services Modernization Act	Allows the combining of commercial and investment banking activities	Dow increased 125% from 1996 to 1999

Figure 4.4   Laws passed to repeal investor protection.

The point to emphasize here is that the security protection laws are passed to protect investors. This almost always happens when people become angry and frightened over scandals and precipitous declines in their portfolios. But when the good times roll, congressmen are pushed to lighten-up on legislation and let the corporations (and investors) do what they feel they need to do in order to rake-in more profits. As often as not, however, this "lightening up" turns a bull market into a full-fledged bubble of the sort we're now sorting out with its likely end being not a whimper but a bang (implosion, actually). More scandals then ensue, more investors get angry and scared, and the whole process repeats itself as is evident from Nofsinger and Kim's study.

As a fitting close to this story of investor protection and government regulation of Wall Street, it's worth noting a remark by Alan Greenspan made in an interview given to the *Financial Times* in July 2008:

The cause of our economic despair, however, is human nature's propensity to sway from fear to euphoria and back, a condition that no economic paradigm has proved capable of suppressing without severe hardship.

Regulation, the alleged effective solution to today's crisis, has never been able to eliminate history's crises.

Let's now leave the businesses of armed conflict and finance, at least for the time being as we'll revisit them in a longer-term context in the next chapter, and examine the trench warfare of commerce. A good place to start is international trade, which might well be termed *war by other means*. Here again we will see how the twists and turns of the global flow of goods and services are dramatically impacted by the beliefs people hold about the future.

LET'S MAKE A DEAL

In July 2003 the United States was busy signing international trade agreements, making pacts and attending summits echoing the peak in global social mood of the time. One deal that clearly reflected the prevailing positive sentiment was a reconciliation between the United States, Germany, France, and Russia. Just four months earlier at the DJIA low, relations were on the verge of a meltdown. But as the mood escalated again, differences over the Iraq War were set aside and trade deals were signed right and left as an outgrowth of the economic summit in summer 2003. But there was a subtle difference between these deals and those signed in the 1990s. The earlier pacts were mainly multilateral in scope; a decade later, the deals were mostly bilateral: The United States signed with Chile and Singapore, China with Hong Kong (!) and India, and India with Egypt. But four months after this flurry of deals, a whiplash turnaround in sentiment is reflected in the collage of media headlines shown in Figure 4.5. The broader story of how the social mood has impacted international trade is displayed in Figure 4.6.

A headline underscoring the shifting social mood's influence on global trade appeared in an Associated Press report of the World Trade Organization talks in Potsdam, Germany, in June 2007. The headline proclaimed, "World Trade Talks Break Up in Failure." The Brazilan Foreign Minister Celso Amorim captured the tone of the meeting when he stated, "It was useless to continue the discussions based on the numbers that were on the table." This comment referred to the United States being criticized by Brazil and India for failing to offer bigger cuts in the subsidies it pays

Figure 4.5 Media headlines about free trade, November 2003
(Source: *Elliott Wave Financial Forecast*, December 2003.).

to American farmers. The USA and the EU in turn said the two BRIC (Brazil, Russia, India, and China) powers refused to open up their markets to opportunities for manufacturing exports. And so it went; hence, the stalemate.

Another sign of the negative mood appeared in a European summit in Brussels in 2007, where the Polish Premier Jaroslaw Kaczynski referred back to World War II in arguing against a per-capita weighting for countries in EU decisions. He said that Poland deserved a greater say because a fifth of its population had been wiped out in the war. The crux of his argument was that "If Poland had not had to live through the years of 1939–45, Poland would be today looking at the demographics of a country of 66 million." Counterfactual history at its best (or worst?).

And what about today? In November 2008 countries representing 90% of world commerce met to yet again publicly acknowledge the destructive nature of trade barriers. At that meeting they promised not to raise trade restrictions for at least one year and to revive the round of trade negotiations begun at Doha, Qatar. In the period since that eventful 2008 meeting, the worldwide negative social mood has begun unravelling those good intentions. Two of the BRIC nations, Russia and India, imposed added import

Figure 4.6 International trade and the DJIA, 1940–2008.

duties, and the US Congress began debating a restrictive "Buy American" clause in financial bailouts. As the weeks and months unfold, anything like a Doha agreement on trade is likely to fade faster than the setting sun in the tropics.

Just to add my bit of sauce to cooking the world-trade goose, here is a collection of trade-oriented headlines I gathered from the *International Herald Tribune* over the 3-month period April–July 2008:

> *In EU, Free Traders are Set to Clash with Protectionists*—April 4, 2008.
> *No Need to Apologize for International Trade*—June 9, 2008.
> *Can These Talks be Saved?*—July 19–20, 2008.
> *Candidates' Dilemma: 2 Sides of Global Trade*—July 23, 2008.
> *World Trade Talks: A Bad Time to Give Up*—July 26–27, 2008.

As Bob Dylan put it, "You don't need a weatherman to know which way the wind blows" when you see an unrelenting stream of stories like this. The global mood has switched from favoring events that would be described as "global" and "joining" to their opposites, "local" and "separating." In the context of international trade, free trade is encouraged when the mood is optimistic, protectionism when it's downbeat. Reflecting the sea change in sentiment since 2007, a recent Op-Ed piece in the *International Herald Tribune* by Daniel M. Price noted that "the former champions of global economic integration have become its biggest skeptics."

So if you're a betting man or woman, don't put much money on the table for passage of a world trade agreement, Doha-style, until the global mood turns back up again. When that will be is a matter we'll take up in the book's final chapter. We now turn to the political arena, where exactly the same psychological processes are at work.

THROW THE BUMS OUT

The political brouhaha in California surrounding the recall of Governor Gray Davis in 2003 is an excellent example of how a general social trend shows itself in a specific political situation. In our opening chapter we mentioned this incident in passing. Now, let's dive-in for a closer look.

At the time, a headline in *The New York Times* stated, "California Mobilizes Recall of Governor Gray Davis." During the bull market, Californians eagerly voted for Governor Davis. He was elected in November 1998, obtaining 58% of the vote. In a September 2002 statewide survey on his handling of economic issues, 45% approved of Mr. Davis and 40% disapproved. He was re-elected Governor in November 2002, capturing 47.4% of the vote versus Bill Simon's $42\frac{1}{2}$%.

But by mid-2003 Davis' popularity as a political leader was at a level last seen in the Nixon era during the Watergate scandal, and a coalition formed with the goal of turning him out of office. A statewide referendum gathered far more than the number of signatures needed for a recall election, and was held on October 7, 2003. Polls at the time indicated that 51% of eligible voters would vote to remove Davis. *The New York Times* of June 16, 2003 quoted James L. Brulte, who headed the state Senate Republican caucus, "It's unbelievable how hated he is," Brulte said referring to Mr. Davis.

What explains this turnaround in Davis' fortunes?

Opponents of Davis cited a list of his bad policies and behavior as reasons why voters turned on him: A massive state budget deficit, mishandling of the state's "deregulation" of electricity, and various other questionable actions and decisions. By way of contrast, *The New York Times* saw no specific actions taken by Davis as causing the revolt. It surmises that "Mr. Davis may be the victim of the increasingly rancorous climate in state and national politics, from the Clinton impeachment to the 2000 presidential election fight in Florida."

This is the wisdom of the conventional analyst speaking. But this type of answer does not account for the dramatic change in sentiment. For example,

- Davis' opponents failed to explain why voters loved him for more than 4 years, re-electing him to office, only to turn upon him suddenly to such a degree that his popularity plummeted to the point of a possible recall. After all, he was the same person then as before, with the same moral character and political philosophy. No change in Davis himself can explain this change in voter attitude.

- One might ask why Davis was re-elected in November 2002 although the social mood as measured by the Dow Jones Industrial Average (DJIA) made a major low on October 9 of that year. The reason is that it takes time for people to mobilize social political actions. And in this case it took about 6 months for the recall movement to gain momentum and visibility. By way of contrast, Richard Nixon was forced to resign right at the 1974 market low with no lag at all. But this resignation took place immediately after the 8-year bear market and right *at the low*. But in the Davis case the bear market had already ended; hence, the time lag.

- An "increasingly rancorous climate" that presumably originated in February 1999 with the Clinton impeachment and simmered in November 2000 over the Florida election battle cannot explain Davis' re-election in 2002 and the subsequent call for his removal just one year later.

A socionomist understands exactly what is going on here. Robert Prechter put it like this,

What a leader does is mostly acausal with respect to the public's opinion of him. There are two reasons for this fact. First, his actions, despite their endless analysis in the press, do little to affect his popularity. Second, his popularity is dependent upon a social mood and economy over which he can exercise no countertrend influence.

As illustration, consider Figure 4.7 showing the public approval rating of one of the luckiest presidents in US history, Mr. William Jefferson Clinton. Despite a sex scandal, which broke publicly in January 1998, and even an attempt to impeach him, Clinton managed to not only retain his presidency but actually *prosper* during his 8 years in office, not at all coincidentally with the fact that the social mood (the DJIA) rose to record levels during this entire period. In the longer-term perspective, history may not deal kindly with Mr. Clinton. But as Henry Ford II put it, History is bunk!

A nationwide survey of American voters in August 2009 reported that 57% would like to replace the *entire* US Congress, while only 25% wanted to keep them (the rest were mixed between all or nothing). This is but a small shift from a similar poll taken in October 2008 when the Congress was passing the unpopular $700 billion bailout plan. At that time 59% of those surveyed wanted to throw the bums out, with 17% opting to keep them. So in a year's time the US Congress managed to win over the hearts and minds of a scant two percentage points of the population. Sounds like a lot of the negative social mood from 2008 hadn't abated at all in the year between the two surveys. And, in fact, a survey in March 2010 showed that

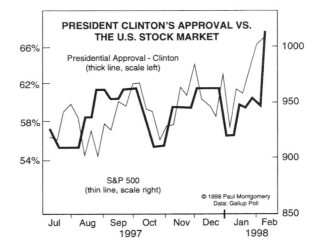

Figure 4.7   Approval rating of President Clinton vs. the S&P 500 index.

14% of the American population believe that Obama is the Antichrist, and 20% say he does the same things that Adolf Hitler did.

To bring things up to date, here are two charts showing the unrelenting decline in popularity of Presidents George W. Bush and Barack Obama. This time the social mood is measured in the *real*, not nominal, stock price, the DJIA measured in ounces of gold. This is to remove the masking effect of inflating away the true popularity levels in a snowstorm of paper currency pouring out of the printing presses in Washington.

The Central Hypothesis of Socionomics implies that an increasingly negative social mood impels voters to magnify politicians' weaknesses, minimize their strengths, perceive that old political styles have failed, and eventually oust the incumbent at the next election or even before it occurs. In early 2003 a bear market had raged for 3 years. Voter hostility had to erupt somewhere. Exit Mr. Davis.

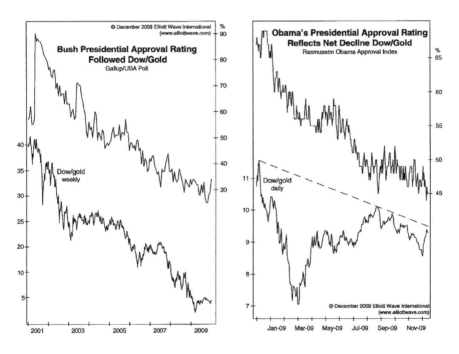

Figure 4.8 Approval ratings of Presidents George W. Bush and Barack Obama.

Earlier, we also looked at the influence voter mood had on US presidential elections, seeing that if the mood is positive immediately prior to the election, the voters without fail re-elect the incumbent or his/her party. What about when we shift from elections to their opposite, what we might term *anti-elections*, things like impeachments and recalls. The situation just described in California leading to the recall of Gray Davis and the election of Arnold Schwarzenegger is a textbook example of what we have in mind.

It's not just in the United States that the social mood biases the political process. Speaking of the June 2009 European Parliamentary elections, the British magazine *The Economist* noted that the European electorate gave "wide support for a ragbag of far-right, populist, anti-EU or plain nutty parties." What's especially relevant about this election is that the parties that were thrown out were not necessarily leaning toward the left. The political orientation of the losers was not nearly as important as the perception on the part of the voters that the party represented the status quo. As *The Socionomist* put it, "In a bear market, the status quo has got to go."

In other parts of the world the same phenomenon is at work. Let me just give a telegraphic overview of what's happened in the two-month period June and July 2009:

- Following the June 12 election in Iran, social tension over perceived fraud in the election gave rise to wave after wave of demonstrations against the incumbents.

- In Peru the worst violence in more than a decade took place in June, as people demonstrated against the current government. These demonstrations led to the death of more than 60 people and the resignation of the country's prime minister.

- Up the road a bit from Lima in Tegucigalpa, Honduras, President Manuel Zelaya was having his own set of difficulties. After Zelaya proposed a modification to the constitution that would have allowed him to run for re-election, the Honduran military had a better idea: a new president, sending Zelaya scurrying into exile in Nicaragua.

All this social turmoil in every corner of the world in just two months! But that's negative social mood for you. Wise politicians will do what they can to get out of the path of this runaway train, in order to live to fight another day. Unwise heads of state like Zelaya get flattened.

What about removal by more systematic and peaceful means than simply the business end of a gun? Different story, same results.

Impeachment is a very rare bird, having taken place only 18 times in US history for Federal officers, basically presidents, members of Congress, cabinet secretaries, and judges. Of those cases there were seven acquittals, each one of which took place at a time of increasingly positive social mood. One of the most recent of those was the acquittal of President Clinton in February 1999 as the mood was moving toward its peak in January 2000. The other 11 cases resulted in resignations, dismissals, and removals from office. Of these, only two took place at a time of increasing social mood, the removal of Judge Halsted Ritter from the bench in the Southern District of Florida in 1936 and the removal of Judge Harry Claiborne in the District of Nevada in 1986. The others were either at a period when the mood was moving sideways or, as in the most recent case, the resignation of Judge Samuel Kent from the bench in the Southern District of Texas in July 2009, were at a time of dramatically declining mood.

The upshot of all these medium-timescale legal and political examples is that when social mood waxes positive, as reflected in persistently rising stock prices, citizens vote to retain the leader who symbolizes their upbeat feelings and that they presume helped cause the positive social mood. When the social mood turns negative, as reflected by persistently falling stock prices, voters want blood, and urge reform laws.

We can summarize the relationship between law/politics and the social mood in the following chart:

Hope	Hubris	Fear	Despair
Relative stability	Maintain status quo	Old rules fail	Radical solutions

Now let's turn away from events that unfold in just a decade or two and take a longer-term view of things. To set the scene, in Chapter 1 we saw in Figure 1.3 that the road to such a long-term phenomemon as globalization began after World War II with the creation of worldwide financial institutions like the World Bank, the International Monetary Fund, and the General

127

Agreement on Tarif and Trade. In the intervening five decades, the global social mood has been basically onward and upward, with only a few short setbacks until it began rolling over for real at the turn of the millenium. Since then we've been on a downhill slide that's gaining momentum as I write. We'll now step back from this precipice and see if a bit of perspective can give an indication as to how long this downhill ride is going to last.

Chapter 5

WHY GREAT POWERS COME AND GO

(Long-timescale Events: A few decades to a few centuries or more)

THE LONG WAVE GOODBYE—AND HELLO

*N*ikolai Kondratiev was a Russian Marxist economist who directed the Conjuncture Institute in Moscow in the 1920s. The principal focus of his institute's work was the study of business cycles in the capitalist economies, since socialist economies do not have the degrees of freedom needed for such cycles to occur. Kondratiev was the first researcher to accumulate a large body of empirical evidence supporting a long-term cycle of about 60 years in things like commodity prices, coal production, interest rates, and wages in western economies, which he published in an article in 1926.

But the temper of the times in the Soviet Union during this period cried out for an actual *theory* of these waves. So in 1928 Kondratiev produced a theoretical explanation based on capital investment. He asserted that capital equipment for production of goods (factories, machinery and so forth) wore out and was replaced in waves of massive investment from savings accumulated during an economic downswing. Thus, while savings is the hallmark of the downswing phase of one of Kondratiev's long waves, increased construction of capital goods is the distinguishing feature of an upswing.

A turn away from capital investment in large projects increases the demand for capital, eventually leading to a rise in the cost of capital. The higher cost then dampens investment leading to the end of the upswing. In the downswing phase, capital becomes increasingly cheaper leading to conditions favorable for the beginning of a new upswing. And so

J.L. Casti, *Mood Matters*, DOI 10.1007/978-3-642-04835-7_5, goes the Kondratiev cycle in 50–60-year oscillations from the upswing to downswing and back again.

Regrettably for Kondratiev, his brand of economics wasn't flavor of the month in the Kremlin, and following publication of his book *The Major Economic Cycles* in 1925 he was dismissed from his post in 1928 and sentenced to a long term of hard labor in the Soviet Gulag. In 1938 he was given the death penalty and executed for his perceived criticism of Stalin's management of the Soviet economy.

It's of historical interest to observe that one of Kondratiev's most vocal critics was the Marxist Leon Trotsky, who characterized long waves as historical periods of accelerated and retarded growth in the development of capitalism. He was claiming, in effect, that long waves are not inherent in capitalism but are a consequence of factors *outside* the capitalist system. External forces, redux!

In Kondratiev's theory there are four cycles, often labeled with the names of the four seasons: *Spring* (improvement), *Summer* (prosperity), *Fall* (recession), and *Winter* (depression). According to aficionados of Kondratiev cycles, each "K-wave" is postulated to last about 50–60 years. Thus, since the beginning of the Industrial Revolution in 1771, there have been four K-waves, the present wave (Summer) having begun in 1949 with an estimated saturation point in . . . 2010.

We bring up Kondtratiev since the idea of cycles of history forms an integral part of the material we'll discuss in this chapter. As our standard of measurement of social mood is a financial market index, it's of considerable interest to see how Kondratiev's cycles match-up to such an index. Figure 5.1 displays a chart comparing the Standard & Poor's 500 Index (S&P 500) during the thirty-year period 1920–1950 (one-half of a full K-cycle) with the same index over the period 1990–2009. In January 2008, Robert Prechter applied this analogy in real time to help identify the start of a major downswing in the stock market.

So what does this have to do with Kondratiev? Looking at these two S&P curves we see a strong similarity between the two, with the top curve peaking just after the high of the K-wave cycle in 1929. The next K-wave cycle peaked around 2000, just where the bottom curve peaks. Moreover, the 1929 K-wave peak was followed by a low in the S&P 500 in 1932, which by analogy matches up with the low in 2002. In the Depression era there was then a high in 1937, which in current times was matched by the S&P 500 high in 2007. The earlier period then saw a downtrend ending

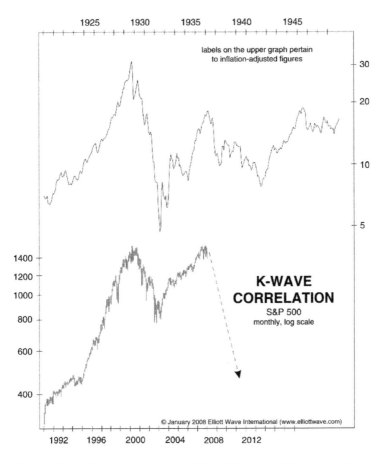

Figure 5.1 Comparison of the S&P 500 at two different periods.

only 5 years later in 1942. So we have both curves peaking along with the corresponding Kondratiev cycles, which lends additional credibility to the analogy between the two periods, 1920–1937 and 1990–2007. Following this analogy, we might expect the current downswing to end in about 2012± 2 years, which is 5 years after the all-time high in 2007. Again, the point is that the end of the most recent Kondratiev cycle led to a pattern in the financial markets that matches the one from the previous K-wave end and is being traced out again as this book goes to press.

OK, you've gotten the idea by now that cycles in both the macroeconomy à la Kondratiev and the world of finance à la the DJIA/S&P 500 give a peek into what's likely to happen next. The balance of this chapter addresses this cyclical pattern in a variety of contexts. To begin, let's return to the battlefield.

WAR AND PIECES

According to many commentators, World War III began at 8:46 AM on September 11, 2001 as American Airlines Flight 11 from Boston bound for Los Angeles was hijacked and crashed into the North Tower of the World Trade Center. History will be the judge of whether this forecast is borne out. Robert Prechter first proposed the idea illustrated in Figure 5.2, which lends credence to the idea. It shows that major wars tend to break out shortly after a major change in trend in the overall social mood (i.e., stock prices). Stock prices were in a downturn in September 2001 that reversed one year later in October 2002. So the 9/11 attack would certainly be consistent

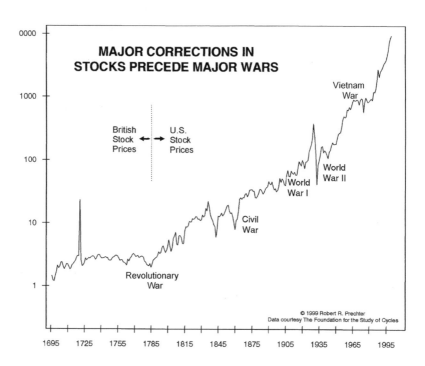

Figure 5.2 Stock prices and the outbreak of major wars since 1700.

with the idea of a major war breaking out after a trend in social mood rolls over.

What Figure 5.2 is trying to tell us can be most compactly summarized in the following table:

Hope	Hubris	Fear	Despair
Old wars end	Little conflict	Increased conflict	New wars begin

If World War III really did begin on September 11, 2001, it's roots can almost surely be traced back to the centuries-long struggle in the Middle East. Can there be a longer-running soap opera on the international stage than this? The battle in its current form began very shortly before the Crash of 1929, when at the end of August of that year a series of what seemed like innocuous disputes involving rights of worship for Jews and Muslims broke out into major bloodshed.

Social clashes take myriad forms, but in 2001 *The Elliott Wave Financial Forecast* identified tension in the Middle East as a "bellwether rift that has an almost perfect record of erupting into open hostility right at the onset of major downturns." Figure 5.3 shows that virtually every major change in social mood was presaged by either the eruption of active hostilities in the Middle East or by a period of easing of tension and hopes for a better, or at least more peaceful, future. (Note: On the chart, the vertical scale is the DJIA adjusted for inflation via the consumer price index.)

A couple of points on this diagram are of special interest. We see that the longest period of goodwill and cooperation between the warring factions was a Middle East *Era of Good Feelings*, which ran from September 1993 though January 11, 2000—*3 days* before a major high in the DJIA. By the summer of 2000 tensions were rising, as the Palestinians threatened to declare statehood. As the social mood declined through the fall of 2000, riots were commonplace and Yassir Arafat's popularity underwent a major shift upward. By May 2001 newspaper headlines were asking, "Is It War Yet?" and Israel employed US warplanes against the Palestinians for the first time since the 1967 war. As stock prices were in a downturn in September 2001 that would only reverse one year later in October 2002, it

Figure 5.3 Ups and downs in the Middle East, 1925–2000.

would certainly be consistent with the idea of a major war breaking out after a trend change in social mood.

CYCLING THROUGH HISTORY

The long waves postulated by Kondratiev are but one example of more general phenomena involving multi-decade cycles in human affairs. Historians have noted such cyclic behavior not only in the economy, but also in the flow of events affecting entire societies. Such "supercycles" entail nothing less than the emergence and disappearance of civlizations themselves. Many authors have examined the birth, rise, decline, and fall of human societies. It's not our intention here to go into the details of these studies, but only to describe the cyclic patterns that each presents, either implicitly, or much better, explicitly. To showcase these patterns, we now look at three of the most well-chronicled investigations of the cyclic patterns in human societies. These investigations were carried out in the early, middle, and late twentieth century by Oswald Spengler, Arnold Toynbee, and Paul Kennedy, respectively.

OSWALD SPENGLER: DECLINE OF THE WEST (1918)

Oswald Spengler was a high-school teacher in Hamburg, when he received an inheritance in 1911 giving him a measure of financial freedom to pursue the life of an independent scholar. Following a move to the headier intellectual atmosphere of Munich late that year, he was struck by the notion that the events of the day could only be properly seen in global and cultural terms. His vision saw Europe marching to a suicidal war, which would be the opening step in the ultimate demise of European culture. World War I (1914–1918) confirmed his prediction, and cemented the thesis he had created and presented in his monumental volume *The Decline of the West*, a work that some have termed one of the most important books ever written. In light of this assessment, it's of interest to note that Spengler encountered great difficulty in finding a publisher for his book, partly because of the radical nature of the work, partly because of the unsettled, chaotic conditions in Germany in 1917. But by the time of the book's eventual publication in the summer of 1918, the mood of the populace was ready for his thesis, and the book became an immediate critical and commercial success.

Spengler was a middle-class German writing in the midst of World War I. His mindset was one of nostalgia for the death of the Second Reich (the empire of the Kaisers built by Bismarck) and he despised the Weimar Republic. He was what today we'd term a "reactionary conservative," who saw democracy as an evil, not a blessing, and who longed for a strong, unifying leader like Alexander the Great or Napoleon. It should then come as no surprise to know that the Nazis saw in Spengler's work a justification for their belief that they were the best way for Germany to dig out from under the crushing reparations imposed upon it by the Allies following World War I. Interestingly, Spengler himself regarded the Nazis as essentially low-grade criminals far removed from his vision of a "grand unifier," and in his later years was totally alienated from their cause.

The Decline of the West is basically an extended argument that the histories of various cultures follow a similar pattern and that all aspects of a culture—art, politics, science, . . .—have related underlying principles that differ only in detail from one culture to another. Let me adapt an account of the three postulates underpinning Spengler's thesis, as originally presented in more complete form by Keith Stimely in an article cited in the References.

- *The "linear" view of history must be abandoned in favor of the cyclical.* Prior to Spengler's work, history had been regarded as a steady progression from lower to higher forms. Thus, western history was seen as developing progressively from Ancient Greeks to Romans to Medieval to Renaissance to Modern. Spengler argued that this view was simply a product of western humankind's ego, that humankind, especially in the West, believed that everything from the past was structured to allow modern humans to create an even more perfect form to live within. Instead, Spengler claimed that history moved in definite, observable, and unrelated cycles.

- *The cyclical movements of history are movements of what Spengler termed "High Cultures."* Thus, they are not movements of nations, races, or events, but something on an entirely different plane. He identified eight such High Cultures: Indian, Babylonian, Egyptian, Chinese, Mayan-Aztec, Arabian, Greek/Roman, and European/western. Spengler also attached a "prime symbol" to each High Culture. The prime symbol for western culture is the "Faustian Soul," representing the upward reaching for the infinite. Part of the theory outlined in the book is that the prime symbol effects everything in the culture, and expresses itself especially in art. For instance, in the Classical Greek and Roman culture the art forms were sculpture and drama; in western culture the great art form is music. Spengler noted that the Faustian Soul symbol is basically a tragic one, as it represents a reaching for something that is literally and logically unattainable.

- The final postulate underlying Spengler's theory is that *High Cultures are living things, they are "organic,"* and thus must pass through the stages of birth, development, fulfillment, decay, and finally death. As a result, there is a kind of "morphology" of history. The high point of a culture is its fulfillment, or culture, phase, its decline and decay then beginning with a so-called "civilization" phase in which there are drastic social upheavals, continual wars, and constant crises. All these unhappy times are accompanied by the rule of Money, together with its tools, Democracy and the Press. The civilization phase ultimately comes to an end with the overthrow of the dictatorship of money, in which great power settles into the hands of great men. This Age of Caesarism sets the stage for the final Imperialistic phase of civilization, in which the Caesars battle each other for control of the Earth. Thus does a High Culture die and sow the seeds for its successor.

ARNOLD TOYNBEE: A STUDY OF HISTORY (1934–1957)

Just around the time the second volume of Spengler's *magnum opus* was appearing in English, Arnold J. Toynbee was polishing off the manuscript for the first three volumes of his own *magnum opus*, which ultimately ran to 12 thick volumes. Initially inspired by Spengler's cyclical theory of history, Toynbee eventually became unhappy with the idea of a Culture as a living organism going through phases of birth, growth, and death. Toynbee agreed that a Culture does indeed go through such phases, but balked at the notion that the Culture is *truly* a living thing. His view was that the only living organisms were the humans making up the population manifesting the Culture, and that Culture itself was not something imposed on the population but rather something that people impose upon themselves.

Toynbee was also less than enamored of Spengler's notion that each phase of a Culture becoming a Civilization lasted a definite, predictable length of time, usually about the same period for each Civilization. Rather, he felt that the lifetime for each Civilization was characteristic of that Civilization, and much more variable than Spengler claimed.

The foundation of Toynbee's theory of history is that Civilizations do not miraculously emerge out of a Primitive epoch in and of themselves, but rather come about when a population of human beings face a challenge of a scale that previous human social groups were incapable of surmounting. This challenge might be a harsh physical climate, the collapse of a previous social group in the same geographical region, an epidemic disease killing off a large fraction of the population, or anything else that threatens the continued existence of the Civilization.

According to the flow of events in Toynbee's theory, a Civilization's growth would come to an end with a "Time of Troubles" because a creative minority would either fail to deal with a new challenge or simply "go to sleep" by resting on their achievements from the past. Breakdown of the Civilization would then ensue as the masses lose faith in the elite, who would then be forced to employ coercion to impose their will upon the population. The Civilization would then enter its Disintegration phase by the rebelliousness of the oppressed majority.

Toynbee's theory asserts that the existing Civilization then attempts to address the challenge. If the response isn't adequate or the challenge is too great, the Civilization is destroyed; on the other hand, if the response is *too good*, the Civilization is held in a kind of paralysis, which Toynbee termed *arrested*. The only type of Civilization that could mature and grow under Toynbee's theory was one that responds successfully to a challenge, but whose response produced new challenges that could also be successfully addressed. In *The Study of History*, Toynbee identified 23 such Civilizations up to the present time. In the 1930s, as the first half of his work was being published, Toynbee claimed that only five of his Civilizations still existed: Russia, Islam, India, the Far East, and the West.

In 1952 the British Broadcasting Corporation invited Toynbee to be the Reith Lecturer for the year. He gave a series of six lectures on the theme "The World and the West," which were later published as a book under the same title. In the first four lectures, Toynbee reflected on the impact the West had on the other Civilizations mentioned above. The fifth lecture addressed the psychological impact of these types of interactions between Civilizations, while the final lecture drew analogies between the impact of the West on the rest of the world and the impact Hellenic Civilization (Greek and Roman) had on the world 2,000 years earlier.

The conclusion Toynbee drew in these lectures that is of greatest interest for us is the well-supported argument that there is a repeatable pattern to the flow of history, a pattern that can be ascertained and employed to gain insight into the structure and timing of events as a Civilization moves through its appointed phases. In this sense Toynbee is firmly in the tradition inspired by Spengler. Let's conclude our brief tour of cyclic historical patterns by looking at a similar set of arguments marshalled by the contemporary Yale historian Paul Kennedy.

PAUL KENNEDY: THE RISE AND FALL OF THE GREAT POWERS (1988)

Paul Kennedy was a relatively unknown British historian based at Yale University when he exploded onto the intellectual scene in 1988 with his bestselling book *The Rise and Fall of the Great Powers*. From today's perspective it's difficult to see why this book caused such a controversy at the time of its publication, as it advocated the rather unremarkable thesis that Great Powers rise to a point where they become overextended due to their commitments to maintain an empire, and that the military and political costs then needed to defend these commitments drains the strength of the economy and forces the Great Power into an inevitable decline to a not-so-great power ("Fall"). Kennedy's book traces this pattern through the Great Powers that have arisen since 1500: Ming China, the Ottoman Empire, the Mogul Empire, Muscovy, Tokugawa Japan, and finally western Europe, ending with a comparison of these Powers to the United States today.

The essence of Kennedy's argument as to why western Europe became the commercial and military leader in world affairs and not, say, the Ottoman Turks or the Ming Chinese is that there was a dynamic involved driven mostly by economic and technological advances, a combination of economic laissez faire, political and military pluralism, and intellectual liberty that were not present in any of the Empires other than the European. As a result, western Europe advanced to the center of the world stage leaving the others in the wings. In short, it was not so much the positive elements but rather the absence, or at least reduction, of impediments checking economic growth and political diversity that gave Europe the edge.

The principle controversy surrounding Kennedy's argument came when he attempted to apply the lessons learned from these Empires to present-day America. The following quote from Kennedy's book summarizes his position:

> ... [America], like Imperial Spain around 1600 or the British Empire around 1900, is the inheritor of a vast array of strategic commitments which had been made decades earlier, when the nation's political, economic, and military capacity to influence world affairs seemed so much more assured. In consequence, the United States now runs the risk, so familiar to historians of the rise and fall of previous Great Powers, of what might roughly be called "imperial overstretch": that is to say, decision-makers in Washington must face the awkward and enduring fact that the sum total of the United States' global interests and obligations is nowadays far larger than the country's power to defend them all simultaneously.

Critics argue that the "rules" of Great Powers do not apply to America, drawing attention, for example, to the historical reluctance of the USA to intervene in the affairs of other nations unless there is a perceived or real threat to American life, as well as the misperception that America was at peace during the Cold War. The only point we need note here is that no one contests Kennedy's facts, and those facts support the thesis that nations rise and fall in a cyclical manner. In short, there is a pattern not only to the ebb and flow of events, but also to the movement of nations to and from center stage in world affairs. And all of the historians we have discussed—Spengler, Toynbee, Kennedy, along with numerous others—have argued long and eloquently that these long-term cycles are not figments of over-active imaginations but are as real as the chair you're sitting on or the book you're holding in your hand. As the social mood shifts, so do a nation's priorities and view of itself and its relationship to other nations. The chart below summarizes the situation from our social mood perspective.

Hope	Hubris	Fear	Despair
Stability	New structures	Old ways fail	Old structures disappear

Now let's turn our attention to some specific examples of these cycles in action.

THE DECLINE AND FALL OF THE EUROPEAN UNION

Changes involving entire nations or even groups of nations are the stuff of which long-term social events are made. Wars, economic restructuring, shifts in political ideologies, and all the things historians love to ponder are for the most part of this type. These are primarily generational changes, unfolding over many decades. It's of considerable interest to see how certain such global transformational events began appearing just around the turn of the millennium in the year 2000. Here we examine just a few of these long-term shifts to get the general picture of what seems likely to be unfolding now and over the next several decades in the geopolitical arena.

May 29, 2005 may go down in the history of the twenty-first century as its first major geopolitical turning point. On that day, French voters rejected the proposed European Union (EU) Constitution, a decision that may well be seen as the last piece in a mosaic marking the twenty-first as the "Asian century." As a consequence of French pessimism about themselves, their present economic situation, and future opportunities, the *Non* vote seemed to be leading Europe to sit by and watch as the world sails on without it. This is a graphic illustration of how the social mood in one region shows the old ways failing, leading to a EU that will probably be like a museum of old cultures and the "good life." According to some commentators, the firm rejection of the EU Constitution by French, and subsequently by Dutch, voters will ultimately send Europe into a state of disarray and confusion, possibly even paralysis, for decades to come. This situation was further acerbated on June 13, 2008 when voters in an Irish referendum soundly rejected the proposed Lisbon Treaty for reorganization of the EU. Of course, the Irish reversed that vote in October 2009 in the wake of a fear of being economically "punished" by Brussels if they didn't "get with the program" during a time of enormous financial stress and disarray in Dublin.

In retrospect, it's easy enough to see that since its founding with the Treaty of Rome in 1957 the EU has never been much more than an economic "marriage of convenience," having little, if anything, to do with a union of cultures, languages, customs, and habits—the very things that typically unify a country or region. A socionomist would recognize this arrangement immediately as one likely to prosper only in times of positive social mood when the economic pie is big enough—and continuing to grow—so that everyone gets an increasingly large share.

Figure 5.4, reproduced from a December 2006 study by Peter Kendall, shows how barriers to unification slipped away as the Dow Jones Euro Stoxx Index reached its all-time high, re-appeared at the initial bear market low in March 2003, and then returned to a lower peak in late 2006. Not shown is the renewed plunge which came as bearish forces were reassuming control over the "expansionary forces" of the 1990s. But as soon as the mood shifts, which it certainly began to do in 2000, the handwriting was on the wall for the EU. The only problem was that the bureaucrats in Brussels and other national capitals didn't seem to be able (or willing) to read the script. And who would have thought otherwise? The French, Dutch, and Irish citizenry gave these bureaucrats the first in what is likely to be a long, painful series of lessons in *Realpolitik*.

Since those fateful ballots, history has not been kind to Brussels. In fact, the city of Brussels itself has been the center of a remarkable schism dividing Dutch-speaking Flanders from the French-speaking Wallonia that threatens to tear Belgium apart. Following general elections on June 10,

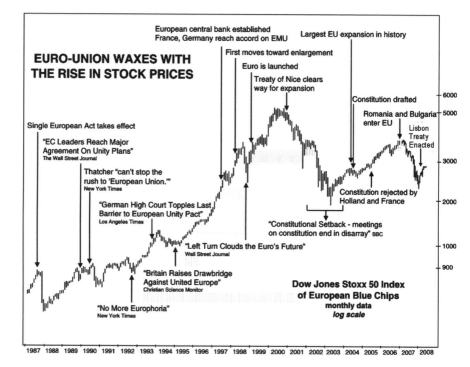

Figure 5.4 Timeline of the European Union, 1987–2007.

2007, Belgian politicians were unable to form a government, throwing the country into a crisis. At that time there was serious talk of the entire nation splitting into two separate states. But calmer heads ultimately prevailed, and on December 30, 2008 a more-or-less stable government was finally sworn in. But this episode, which lasted for more than 18 months, was ample evidence of the power of the negative social mood that's still sweeping Europe to foment old, long-standing divisions and bring them back as forces of separation and fragmentation, just the opposite of the unification theme supposedly underlying the EU.

Columnist Jonah Goldberg of the *Los Angeles Times* wonders if Belgium isn't the first of many fractures in the EU. In an October 9, 2007 column he wrote,

You probably don't realize it, but we are living in an unprecedented historical moment. Belgium is coming apart at the seams. ... If Belgium falls to sectarianism, what does that say about prospects for making Europe into a super-Belgium?

Socionomically speaking, it says that Belgium is leading the way into a new era of Euro *dis*union. As Peter Kendall forecast in the *Elliott Wave Financial Forecast* in 2005, "During the bear market, the independent nations of Europe will rediscover their borders and rekindle the animosities that kept them apart for centuries."

Perhaps the most common explanation given in the media for the French, Dutch, and Irish antipathy to the EU Constitution was a fear of globalization. Loss of jobs, inability to compete in the global marketplace, and outsourcing of work, along with an adamant insistence by unions on retaining "sweetheart" arrangements for workers, involving such things as increasingly shorter work weeks and generous social benefits, were just a few of the reasons pundits offered for the constitutional turndown. Strangely, while the evidence mounts for a dramatic, decades-long shift of economic power from North America and Europe to Asia, there is plenty of evidence also mounting *against* globalization.

In many ways the EU represents a kind of "dress rehearsal" for globalization, starting on its path to the form we see it today with the Treaty of Rome in 1957. I could recount numerous headlines over the past few years describing European concerns such as "Islamophobia" and other tensions over immigration, the right of "free expression," Brussels-style, the Belgian separatist movement, expansion of the EU to "non-European" nations, and the like. The chart we saw in Figure 5.4 summarizes the social mood in Europe. We see the ups and downs of the EU following in lockstep with the social mood: Nice things tend to happen when the mood is positive, not-so-nice things when the mood shifts downward.

For those who prefer to take their moods measured by public opinion surveys, it's worth mentioning the *Eurobarometer* survey of citizens from EU countries who were asked to respond to the statement "Our country's membership of the EU is a good thing." Of the 15 EU countries surveyed, the average percentage of people across Europe who answered in the affirmative in 2006 was just over 50%, up very slightly from about 45% 10 years earlier—not much progress for a period when the social mood was about as positive as one could ever hope for.

Interestingly, at least for Austrians, is the fact that only the UK had a lower percentage of positive respondents than Austria. For the curious, citizens of Ireland and Luxembourg were the most positive. Yes, Ireland! Which only goes to show how fast people's minds can change in the space of a year. In Chapter 1 we gave a telegraphic account of the decline and fall of globalization. Now let's dig a bit deeper into the matter.

According to Canadian philosopher John Ralston Saul, grand economic theories seldom last more than a few decades. He cites unbridled capitalism that died in 1929 after a run of about 30 years, Communism, a blend of anti-religious, economic, and global theories that lasted 45–70 years, depending on geographical regions, Keynesian economic policies that ran for about 45 years, and today's globalization which has been on center stage for roughly 30 years. In Saul's view it's just about time for globalization to start singing its own swan song. Perhaps the situation in the EU is the opening stanza of a very long medley of songs, since grand ideologies seldom leave center stage voluntarily, generally having to be dragged off, kicking and screaming. But the signs of decline are clear—if you're a socionomist—and they are turning a muddled and confused situation into a landslide victory for the anti-globalization forces.

To understand what's happening, here is a brief summary of the arguments presented in Saul's immensely illuminating article "The End of Globalism," which every reader of this book should ponder. It is available on the Internet as noted in the References.

First of all, globalism is based on the notion that the nation-state as a power is passé and can be replaced by the power of the global markets. So economics, not politics, would determine the course of human events. Such global economic forces would protect the populace from the mistakes of prideful, local leaders, and give the individual a better life through enlightened self-interest. Or so goes the theory, anyway.

Historically, one can trace the birth of globalization to 1971, where in the Swiss mountain village of Davos, a club of corporate leaders was founded to look at civilization through the eyes of business. Soon the business people were joined by feckless government leaders and greedy academics looking for sponsors for their very definite, but almost totally meaningless, projects. The ideology that bound this group together was the notion that the public good could be regarded as a by-product of trade, competition, and self-interest. Four years later, the G8 was founded to mimic the Davos ideology that the world should be seen as one great corporation, ignoring social standards, human rights, organized religions, and anything else not overtly aimed at commercial self-interest.

Of special interest is that what really opened the door to globalization was the economic crisis that began with the oil-price shock of 1973. The new ideology that came forth to fill the vacuum left by the leaders who had lost the ability to deal with the situation was called globalization. As with all successful religions, globalization promised to solve every problem. Simple, broad-sweeping solutions with responsibility ultimately lodged in the hard-to-pin-down, invisible hands of corporate moguls. No one had to take responsibility for anything; the markets would miraculously solve all.

For two decades, government after government simply shifted most of their economic power onto the global markets by passing legislation favorable to global corporations, entering into alliances like the EU, enabling floating currency exchange rates, and the like. As the unregulated debt and currency systems allowed transnational corporations to accumulate assets greater than most countries, it was a logical step to regard these firms as new nations themselves, a kind of virtual nation devoid of geographic boundaries, citizens, obligations. Finally, in 1995 the system peaked with the GATT agreement leading to the founding of the World Trade Organization, a central body created to deal with matters of trade. So the reconceptualization of civilization through the looking glass of economics reached its critical point. Socionomically speaking, this happened just as the social mood was moving strongly into positive territory with the Great Asset Mania of the 1990s.

I'll skip past some of the more illuminating and thought-provoking parts of Saul's argument and give only the gist of his main conclusion. Again, I urge the reader to consult the full paper for details of how globalization led to the exact opposite of its proclaimed goals of enhancing individual choice, fostering a worldwide growth in democracy, and reaping the fruits of privatization of everything from airlines to electric power to postal systems.

Following the breakaway policies of many governments in Central and South America to re-take control of their own destiny from the global corporations, in 1998 Malaysia refused to follow the global rules and pulled its currency off the market in response to the economic crisis in Asia at the time. The Malaysians made their currency nonconvertible, and fixed its value at a level just low enough to benefit exports. They also blocked the export of foreign capital and imposed tarifs on imported goods—all to the horror and condemnation of western governments and corporations, who one and all predicted an imminent collapse of the Malaysian economy. But no such thing took place.

Just one year later, in 1999, bankers and other leaders began to praise the Malaysian "experiment." And at the very same time, the citizens of New Zealand, a country that had become almost a poster child for globalization 15 years earlier, voted to change direction by endorsing a strong government hand in regulating both the national economy and the private sector. Again, a socionomist would recognize this as the very time when the social mood was on the precipice just prior to the collapse of the financial markets in January 2000. As the table of moods presented at the end of the previous section suggests, this is just the time when one might expect the "old ways" to begin to fail, and fail noticeably.

As an epitaph to the entire notion of globalization, in January 2003 the "visionaries" of Davos invited Malaysian prime minister Mahathir Mohamad to address them on the opening day of their annual meeting, heaping praise upon him for his country's economic success. All of a sudden the reconstructed globalists realized that the Malaysian success stemmed from leadership at the nation-state level and was based upon a total rejection of globalist economics. The frosting on this particular cake—or the last nail in the coffin, depending on your choice of metaphor—came just a few days later when newly-elected Brazilian president, Luiz Inacio Lula da Silva, laid out a vision of responsible nation-state populism to the same audience. So in a span of just a few days, in the very birthplace of globalization, several governments of very different types all turned away from the concept and painted a picture of the world as if the nation-state were again the pillar upon which power and influence rest.

Earlier, we gave an account of how socionomics would argue that the negative votes on the EU Constitution, with their emphasis on anti-globalization, is simply an event born of the social temper of the times. After all, who wants to back a losing horse—even if it's your own horse? As for the future of the EU, the socionomic take on that issue is crystal clear from the story told above. *Sic transit gloria*!

To get a feel for where we currently stand on the question of global actions, Figure 5.5 shows a spectrum of world events over the last eight decades. Our rule of thumb is that "bullish" sentiment means that things are getting better, while "bearish" indicates a turn for the worse. In this chart there are two basic periods: a bear market during the Great Depression, followed by a bull market that began after World War II. It doesn't take much insight to see the qualitative difference in character between the world events taking place during each of these periods. But it takes a socionomist to recognize the social signs of a juncture at hand and to place the word *top* on such a graph.

Figure 5.5 World events during two supercycle waves, 1920–1999.

Let me close this section with a few questions for the reader: With this bittersweet story of the rise and fall of globalization in mind, where do we now stand on the chart of geopolitics shown above? And where is the international scheme of things heading today? And how does the rise of Southeast Asia as a force impacting the global economy fit into this emerging picture? Finally, what about the course of development that seems to be favored by the chattering classes and talking heads in the media, who focus on the rise of the so-called BRIC countries, Brazil, Russia, India, and China? Twenty years should be about right to see answers emerging to these puzzlers.

BRICS OR BRIC-A-BRAC?

In early 2007 the *Asia Times* noted,

The coining of BRIC was a boon for the emerging markets. The snappy acronym summed up the macroeconomic forces that were reshaping the planet. The acronym sparked myriad new investment funds and trading

147

strategies, boosting capital flows into the emerging markets and trading commissions for investment banks.

Boy did it ever! Between May and September 2008, BRIC stocks sank more than 70% from their peak. And this was in the face of media statements that the BRICs will display "a resilience unimaginable in the U.S." If you believe that, you'll believe in the tooth fairy too. Figure 5.6 shows the situation with the BRICs just a few weeks later.

As if this sentiment wasn't rosy enough to emit the heat needed to warm up the BRICs, the head of one major investment firm stated, "I would expect the big emerging markets to do really well in the updraft of the next bull market, which you ought to be positioned for right now," a statement made in early 2009. So where do things stand today in the biggest BRIC of all, China? The picture as I write is shown in Figure 5.7. The Shanghai Exchange had the same run up as most other markets from late 2008, but

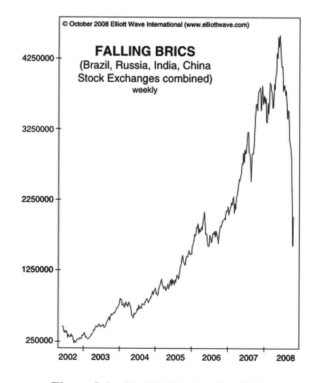

Figure 5.6   The BRICs, October 2008.

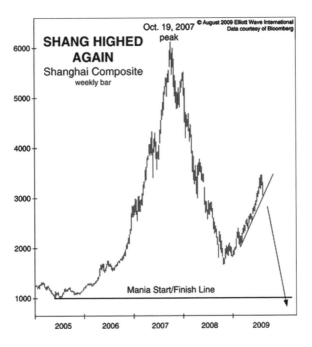

Figure 5.7   The Shanghai index thru August 2009.

as with the rest of the world it's in the final stages of that bear-market rally, destined ultimately to drop down below the "starting line" of the big surge upward. So maybe bric-a-brac is a more suitable term for these emerging markets, after all.

Perhaps, though, China is not the poster child for the BRICs anyway. Its position in the overall global scheme of things suggests that it should not be lumped together with the others. In fact, to preserve the acronym substitution of China by Canada would be more appropriate.

There's no room in a book of this type to go into detailed analysis of all the BRIC countries, so let me choose just one of the BRICs to look at in more detail. Since China is by far the most well chronicled of the four, as a European resident I'll indulge my personal interests for the next few pages and focus attention on the political, economic, and military situation in Russia as it relates to the ever-shifting social mood.

A good place to start a study of Russia is at the beginning of the long-term decline that began around the year 2000. From 1998 to 2006, Russia's

149

economy grew at the furious pace of 6.7% annually. By way of comparison, in 2006 Germany's growth rate was 2.7%, France logged-in at 2.2%, and The Netherlands registered 3%. In fact, in Eurasia only China, with its 10.7% annual growth, was larger.

To go along with these growth rates, Russia increased its foreign reserves from $12 billion to $420 billion since 1999, the third largest in the world behind China and Japan. And as investment money flowed into Russia, household consumption grew at a staggering pace along with it. Personal loans rose by 75% in 2006, while outstanding mortgages rose more than 460% (!) from 2006 to 2007.

As for changes in the Russian power structure, Peter Kendall has noted that over the last 80 years Russian power grabs appear not-so-myteriously at peaks in social mood. For instance, in 1929 Stalin effectively seized the Russian government with the expulsion of Bukharin and Rykov from the Politburo and the exile of Trotsky to Mexico. At the next major peak in 1937, the Great Purge began, during which Bukharin and Rykov were executed. The next important peak came in early 1966, a time when Brezhnev came to power. The Brezhnev era was marked by a more repressive, authoritarian tone and took until the early 1970s to consolidate.

It seems as if now Putin is making a similar transition. His first act as president was to take control of the Chechen capital of Grozny on February 6, 2000—right between a peak in the DJIA on January 14 and in the S&P 500 on March 23. As for what's next, Alan Hall of The Socionomics Institute notes,

> Putin's latest power grab is probably right up there with Brezhnev in the mid-1960s. "To become the head of government is quite a realistic proposal," Putin says without mentioning that it means abandoning the intent if not the letter of the Russian Constitution of 1993. According to Bloomberg: "The plan allows Putin, 54, to continue exercising control over the nation while adhering to the form, if not the substance, of a democratic transition."

And when did this latest power grab occur? In autumn 2007, just as the social mood (i.e., the DJIA) was reaching its all-time peak!

What this picture portrays is that positive social mood is a wondrous thing. And it's also a powerful engine for growth, changing Russia from a feeble democracy to a major global financial power in the space of just

© November 2007 Elliott Wave International (www.elliottwave.com)

Russian Economic Boom
—VOA News, Oct. 29, 2007

Putin: 85% approval
—Angus Reid, Jul. 22, 2007

RTSI$/gold

**THE RUSSIAN AND
U.S. STOCK MARKETS
IN TERMS OF REAL MONEY**

DJI/gold

Bush: 34% approval, Congress: 22%
—MSNBC, Nov. 5, 2007

"A nation of discontent"
—USA Today, Nov. 1, 2007

Figure 5.8   Russian social mood vs. US, 1999–2007.

a few short years. This positive mood is seen graphically in Figure 5.8, which shows the diverging social mood in Russia and the USA measured by the two countries' respective stock indexes denominated in *real* money, gold, rather than the paper "Monopoly money" cranked out by government printing presses in Washington, DC and Moscow.

By the end of the 8-year period 1999–2007, Russian belief in the future had expressed itself in many ways:

- Real-estate fever struck Moscow like in no other city in the world. As an example, the city fathers initiated a plan to sell Gorky Park, the Moscow version of Central Park in New York City, to a developer who hoped to create ultra high-end apartments in this prime location selling for $10,000–$20,000 per square meter (a cool $1–$2 million for a 1,000 square-foot flat).

- In August 2007, the Russian and European space agencies announced plans to jointly develop spacecraft for flights to the International Space Station, the Moon, and Mars. Moreover, Russia became the first country to offer "space tourism" to well-heeled customers ready to fork over millions for a trip to the space station.

- A Russian explorer, Artur Chilingarov, laid claim to the North Pole and half the Arctic (including all oil and gas deposits there) in August 2007, when he planted the Russian flag on the seafloor beneath the Arctic Ocean. Both the USA and Canada disputed this claim, which led Chilingarov to say he didn't give a damn about the politicians, and if others didn't like his claim they should go to the seafloor and plant their own flag. In closing an interview with CNN, Chilingarov stated: "The Arctic has always been Russian."

- At the end of 2007, the Russian public gave President Vladimir Putin a nearly 80% approval rating. Putin's image, in sharp contrast to that of his predecessor, Boris Yeltsin, who was perceived as a clownish drunk, was that of a serious, sober, and energetic "doer." As a manifestation of this image, the Russian media delighted in presenting photos of a bare-chested Putin high-stepping through a mountain stream on his summer holiday in Siberia. The boulevard press newspaper *Komsomolskaya Pravda* reported that women visiting its web site were "screaming with delight and showering him [Putin] with compliments." And Russian radio plays a pop song titled "A Man Like Putin," in which the singer voices her desire to leave her drunken boyfriend and link-up with "Someone like Putin, full of strength; Someone like Putin, who doesn't drink; Someone like Putin, who doesn't hurt me; Someone like Putin, who won't run away." Such uncritical approval comes only at extremes of positive social mood.

These small vignettes are from Hall's special report, "Sizing Up a Superpower: A Socionomic Study of Russia," which appeared in late 2007. Together they show how the extreme of positive social mood in Russia at the time was translated into a flexing of the Motherland's economic and political muscles on the world stage. But extreme presidential popularity tends to coincide with extremes in social mood along with everything else. On the basis of Putin's unrivaled stature within Russia, as well as various other socionomic signs of a peak, Hall called for an eventual decline of at least 50% in Russian stocks. So what's happened to Russia since then in the wake of the Great Economic Crisis? Let's go to the data and find out.

The chart in Figure 5.9 shows the social mood in Russia (the index of the Russian stock market) during the Putin era. Since the extreme of positive social mood in May 2008, the index has blown-off nearly 60% of its value through summer 2009. And, in fact, if one counts to the low in social mood in spring 2009, the picture is even worse—a nearly 80% falloff in social mood. That's pretty impressive in just one year. Talk about a reversal of fortune. This was more like whiplash than a mere turnaround!

Of course, this change of social mood has unfolded on a timescale of many years, somewhere between what we have termed *medium term* to *long term*. Thus, from a day-to-day perspective it may appear broad and slow, even glacial to some eyes. Hall has noted a spectrum of news stories from the late 2007 period that gives some indication of the coming shift in Russia from emphasis on the economic back to the military.

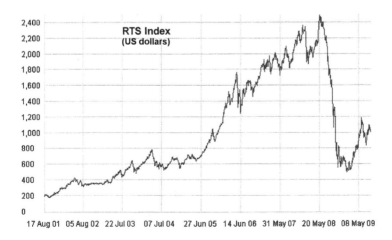

Figure 5.9 Russian Trading System index, 2001–2009.

In late 2008, the *Rossiya* TV station ran a survey among their viewers of who they thought were the greatest Russians in history. Top place went to Alexander Nevsky, a medieval Russian king who defeated crusaders from Germany and Sweden in a battle on the frozen Neva River in the 1200s. But the great surprise was the person who came in third: mass-murdering dictator Joseph Stalin. And at one point he was leading, only to end up a very close third (behind Nevsky and Piotr Stolypin, who tried to modernize Russia after the fall of Czar Nicholas II in 1917 but was assassinated before he could finish the job). Stalin actually ended up trailing Nevsky by a scant 5,000 votes in a poll in which several million votes were cast. How could such a tyrant responsible for the deaths of millions of Russians receive over half a million votes?

Many commentators point to Stalin's leadership in World War II and the fact that Russia maintained a strong position of power during his further reign as the source of his popularity today. As a surface explanation this is as good as any, I guess. But the deeper reason is that at the time of the *Rossiya* survey, Russians were (and are) mired in a period of increasingly negative social mood. And when people are fearful of the future, they long for a strong leader and wax nostalgic over a past in which they felt confident and secure in their future. As stated by Oscar-winning Russian film director Nikita Mikhalkov, "We need to think very deeply about the fact that Josef Vissarianovich Stalin came third. This is a cause for us to ask what the mood of the people is, what is it they want."

But perhaps this poll isn't to be taken at face value, since a similar poll carried out in the UK in 2002 declared Princess Diana the third greatest Briton of all time, just behind Stalin's old sparring partner Winston Churchill and Isambard Kingdom Brunel.

In a more serious vein, though, Vladimir Putin lent his authority to the poll even earlier when he endorsed a new guide for history teachers, *A Modern History of Russia: 1945–2006*, which describes Stalin as "One of the most successful leaders of the U.S.S.R," who created "the best educational system in the world." Later, this revisionist history describes the results of Stalin's purges as "A new class of managers capable of solving the task of modernization in conditions of shortages of resources, . . . and immaculate from the point of view of executive discipline." And it's not just a resurrection of Stalin. The new/old Russia has been flexing its military might under the Putin regime, as well. Here are a few straws in the wind illustrating this trend.

Item: Bombers—On August 8, 2007 Russian bomber pilots exchanged smiles with US fighter pilots who met them above a US naval base on Guam. But smiles were not the order of the day a month later when British forces scrambled four fighters to intercept eight Russian bombers over international waters. According to Reuters, "Relations between London and Moscow are at their worst since the Cold War."

Item: Missiles—Russia's military budget has quadrupuled since Putin took power, with considerable attention being paid to a new generation of nuclear submarines and nuclear missiles capable of evading US anti-missile systems. Russia claims this heightened military posture is due to US withdrawal from the Anti-Ballistic Missile Treaty and American plans for a European-based missile shield.

Item: Ships—The Russian navy has returned to patrol the Mediterranean Sea and began building docking stations in Syrian ports. Russia claims that this deployment is purely symbolic.

Item: Protectionism—Western oil interests of Exxon/Mobil, British Petroleum, and Shell have taken it on the chin from the Russians, as the "Putinocracy" snaps up gas and oil fields right and left under the pretense of violation of safety standards and tax evasion by the western firms. For example, Exxon/Mobil developed the Sakhalin 1 gas fields. But when they were ready to start exporting the gas to China to recoup their investments, the Russian gas monopoly Gazprom told Exxon not to honor the contract with the Chinese because the gas was needed at home in Russia—at a vastly lower price.

On and on it goes: human rights violations, massive internal corruption, judicial sleight-of-hand, belligerent rhetoric, increasing authoritarianism, declining population, suppression of the media, nationalistic youth movements, to name but a few manifestations of post-2007 Russia. Just like the good/bad old days!

To conclude this list of semi-abstract characteristics, here is a sampling of headlines taken from the *International Herald Tribune* newspaper over the past few months:

Russia's Resurgence: Risks and Rewards—July 26, 2008.
Churchill's Definition of Russia Still Rings True—August 2–3, 2008.
For Russia's Oligarchs, What Price Georgia?—September 4, 2008.
Russia Sees In Credit Crisis the End of US Domination—October 3, 2008.
In Russia, Fear that Financial Turmoil is Back—December 30, 2008.
Another War? Russia's Next Flashpoint—January 7, 2009.

So what does this all add up to? I cannot do better than to cite the closing statement by Michael McFaul and Kathryn Stoner-Weiss in their *Foreign Affairs* article examining the pros and cons of Putin's authoritarian model of governance in Russia. They conclude with the following statement:

> The Kremlin talks about creating the next China, but Russia's path is more likely to be something like that of Angola—an oil-dependent state that is growing now because of high oil prices but has floundered in the past when oil prices were low and whose leaders seem more intent on maintaining themselves in office to control oil revenues and other rents than on providing public goods and services to a beleaguered population. Unfortunately, as Angola's president, José Eduardo dos Santos, has demonstrated by his three decades in power, even poorly performing autocracies can last a long, long time.

A fitting testament to the power of a rapidly-declining social mood to undo everything in a country that the previous wave of positive mood created—and more.

WHAT ABOUT AMERICA?

In 2009, *Slate* magazine ran a feature in which they asked their readers to examine 144 potential apocalyptic scenarios for the end of America and choose up to five that taken together they felt described the most likely routes for the United States to disappear from the map. These scenarios ran the gamut from a supervolcano eruption (probably in Yellowstone National Park) to hyperinflation to globalization and world government. More than 60,000 readers responded, many from outside the United States itself, including a healthy 15% from Russia. More than a quarter million individual five-event scenarios were selected out of the more than 3 million combinations of five events chosen from 144, suggesting that there was great variation among the choices. The five most "popular" scenarios and the fraction of respondent who chose them were as follows:

- *Loose Nukes (10.5%)*—a nuclear attack on the United States by terrorists of one stripe or another.

- *Peak Oil (9.3%)*—petroleum production falls off the cliff, as oil becomes too expensive to extract and alternative energy sources don't pick up the slack, sending the USA back to a horse-and-buggy style of life.

- *Antibiotic Resistance (8.5%)*—the widespread sale of antibacterial hand-soaps sparks the development of superstrains of bacteria resistant to all known antibiotics.

- *China Unloads US Treasuries (8.2%)*—the United States is brought to its knees through an economic collapse triggered by China pulling the plug on the American economy.

- *Arab-Israeli War (7.6%)*—the long-awaited denouement of the festering tensions in the Middle East finally goes "hot," drawing in the United States on the side of Israel. This, in turn, leads to widespread global hatred of America and ultimately destruction of the American lifestyle by a combination of terrorism and economic collapse.

So these are the top five ways for the USA to quickly vanish from the spotlight on the world's geopolitical stage. Just for the sake of balance, let me mention the least popular scenarios: An Anthrax Attack (0.5%), Pax Indica (0.6%), and Declining Military Standards (0.6%).

I won't go into the many interesting details of this survey, such as why men were far more concerned about China than women, as the reader can find a detailed discussion of these matters in the article cited in the References. But one thing that's worth quoting here is the most popular *grouping* of scenarios, as opposed to the most popular individual scenarios. The most popular group appeared 102 times, and according to the editors leads to the following pattern of events:

First, China pulls away from the United States, sending inflation spiral-ing out of control and dropping America from the ranks of economic superpowers. The economic crisis in America causes a global recession, leading to spikes of terrorist activity worldwide. Meanwhile, petroleum production slows down, and water supplies begin to dwindle. This causes yet more instability, and global terror networks—which blame the West for the world's decay—take advantage by filching nuclear devices as unsteady regimes collapse. On the darkest day in human history, the

United States gets strafed by nuclear bombs, killing millions upon millions of American citizens. Those who don't die during the horrific nuclear attack get finished off by antibiotic-resistant bacteria that public-health officials are unable to get under control. The End.

For readers who'd like to write their own end-of-America apocalyptic fiction, here's the second most popular apocalypse grouping, chosen 52 times: Loose Nukes, Russia Hits the Button, Invasion by a Foreign Power, Electromagnetic Pulse, and Dirty Bombs.

At this point the reader might be wondering what all this apocalyptica has to do with social mood and events. Hopefully, that same perceptive reader will recognize these end-of-America scenarios as something whose likelihood shoots way up during a period of negative global social mood. And by now it should be manifestly clear that this is the type of period the world will be facing for the next several years, if not decades. With this background in mind, perhaps these doomsday scenarios are not so farfetched after all.

But what about a less dramatic, and probably more realistic, set of possibilities than total destruction? A bit of historical perspective on this issue is provided by Figure 5.10, which shows some major events in the history of the United States from 1760–2000. For future reference, a good

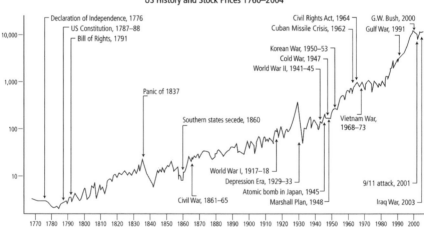

Figure 5.10   US timeline, 1760–2000.

source of inspiration is a report produced by the US National Intelligence Council in 2008 titled *Global Trends 2025: A Transformed World*. Here are a few highlights from this extremely interesting and thought-provoking document.

- The world is in an unprecedented "transfer of wealth and power" from West to East (mostly Asia). The drivers of this transfer are the increase in the prices of oil and other commodities, along with the shift of manufacturing from the industrialized western countries to Asia.

- The international balance of power established after World War II will be changed to an almost unrecognizable degree by the rise of new powers, a globalized economy, a huge transfer of wealth from West to East, and the growing influence of non-state actors.

- America's biggest rival by 2025 will be China, which will have the world's second largest, or possibly largest, economy by then and will be a major military power as well.

- Other countries such as Iran, India, Indonesia, and Turkey will also see their power increase, along with a need for natural resources.

- American power will decrease relatively speaking from where it is today, although America's obligations and burdens will not. These include the jobs of world "policeman," "global banker," and "consumer of first resort."

The bottom line as envisioned by the authors of the report is that "The next 20 years of transition to a new system are fraught with risks." To be honest, this isn't much of a surprise—or much of a conclusion. And it's especially not surprising to a socionomist, as all the points above are foreshadowed by the steadily shifting mood from one of "global," "joining," and "welcoming" sorts of events to those of a "localizing," "separating," and "xenophobic" cast.

The cases we've examined in this chapter have thus far been "long-term" in a kind of short-range sense, mostly ranging over a few decades with the exception of the long waves in the rise and fall of civilizations. So let's close the chapter by considering some seriously long-term phenomena, those extending over millennia.

CHAPTER 5

THE *REALLY* LONG VIEW

The 2005 Atlantic hurricane season was the most active in recorded history, repeatedly shattering previous records. The impact of the season was widespread, expensive, and deadly with at least 2,280 deaths and recorded damage of over $128 billion. The most catastrophic effects of the season were felt on the United States' Gulf Coast, where a 30-foot storm surge from Hurricane Katrina caused devastating flooding inundating New Orleans and destroying the majority of structures on the banks of the Mississippi. By late summer, residents of Florida and the Gulf Coast were battening down the hatches to protect their property and their lives from the rage of these forecasted hurricanes. This type of precaution is a good metaphor for the world situation in the coming years, at least if one takes the socionomics perspective on how the global social mood is unfolding. First the bad news.

Just as hurricanes come around in a predictable cycle, even though specific timing, landfall locations, and intensities of these storms are a matter of probabilities, not certainties, the social mood of a population also rises and falls in accordance with a pattern that is every bit as predictable—and unpredictable—as the vagaries of hurricanes. Both are law-like phenomena, one governed by the laws of nature, the other governed by the laws of human nature. And to complain about either or to imagine that you can somehow escape these laws is about as useful as complaining about the Sun rising in the east and setting in the west each day. Maybe one day that solar pattern will be reversed. And maybe one day basic human nature will change, too. But neither one is the way to bet.

The long-term transition from a positive social mood to the negative one the world is now just entering is a consequence of the biological nature of humankind that's hardwired into the deepest structure of the human brain and created by the social nature of human beings. So if the social mood is truly turning negative at the long-term level, then the world is headed for a downturn that, to paraphrase Thomas Hobbes, will be nasty and brutish—but unfortunately not short.

To get a truly long-term perspective on this fact, Figure 5.11 shows the wave pattern characterizing the social mood over a period of two *millennia*. You might ask where does one get financial data to infer the social mood centuries into the past? After all, there was no New York Stock Exchange or DJIA in ancient Rome. The simple answer is, you don't. From about 1790 onward good financial data is available. But for earlier times one must

MILLENNIUM WAVES

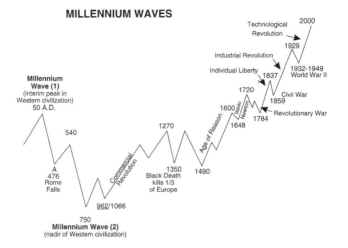

Figure 5.11 Millennia wave pattern for the past 2,000 years (Adapted from Fig. 18.7 in Prechter, R. *The Wave Principle of Human Behavior and the New Science of Socionomics*, 1999.).

"reverse engineer" the situation. That means examining the nature and character of historical events and placing them as to the type of social mood that would be most likely to give rise to conditions favoring such an event. Robert Prechter carried out this procedure for numerous events of all types already written onto the historical record, coming up with the chart above. What's remarkable is that although Figure 5.11 is only a crude approximation to general price trends, what we see is a very long-term postive mood (bull market) beginning at the end of the Dark Ages—a bull market we've been in ever since!

This observation opens up the question of the "millennium" bear market that preceded this bull market. The best guess is that it began shortly after the death of Christ and continued for several centuries of misery—disease, slaughter, and stagnant social progress. A deeper discussion of the construction and implication of this diagram is found in the books cited in the References. For us, the most important feature is the strong suggestion that the period of positive social mood that began around the year 1000 AD is now coming to a close. The millennium wave is still in its infancy, so we can expect it to last *at least* another 1,000 years and probably considerably longer. So at the societal level we'd better get used to the idea.

The good news, though, is that while the societal level is one thing, the timescale for individuals is something else again. In the final chapter we take up the matter of how to use the social-mood-based ideas identified and explicated in the preceding chapters to get some insight into our future as individuals, as well as how to squeeze a bit of short-term optimism out of the dominant trend downward. This closing chapter distills the essence of the myriad ideas and examples of the foregoing pages into a workable procedure for looking into the future and somehow coming to terms with it.

Chapter 6

GETTING IT TOGETHER

How to Anticipate the Future

POLISHING THE CRYSTAL BALL

*I*n 1956 the American Petroleum Institute held its annual meeting in San Antonio, Texas. On the program was a talk titled "Nuclear Energy and the Fossils Fuels" to be presented by M. King Hubbert, a geophysicist at the Shell Oil Research Center in Houston. While attendees at the meeting had little inkling of what lay behind this rather bland title, Hubbert's employers certainly did. The head office of Shell was on the phone to Hubbert virtually up to the minute before he strode to the podium, begging, threatening, and cajoling him to withdraw the presentation. But if nothing else, Hubbert was a stubborn old cuss who stood behind his work, and thus ignored these pleas and unleashed what has today come to be called the "Peak Oil Question," or more informally, "Hubbert's Peak."

What Hubbert argued was that US oil production would peak in the early 1970s, a claim that no one in the oil industry wanted to hear then—or now. Figure 6.1 shows Hubbert's argument graphically. The upper dashed line in the figure, labeled for ultimate reserves of 200 billion barrels of oil, is what got everyone's attention. In fact, actual US production from 1956 to 2000 was a bit greater as shown by the small dots at the top of figure. This is due to production from Alaska and far-offshore fields in the Gulf Coast. But the difference is small, and the general drift of production is following Hubbert's curve perfectly. Today, the whole question of Peak Oil has been moved up a level, with a recent Kuwaiti study claiming that *global* oil production will peak in about the year 2014. If these predictions are even close to being as accurate as Hubbert's were for US production, the world

J.L. Casti, *Mood Matters*, DOI 10.1007/978-3-642-04835-7_6,   163
© Springer Science+Business Media, LLC 2010

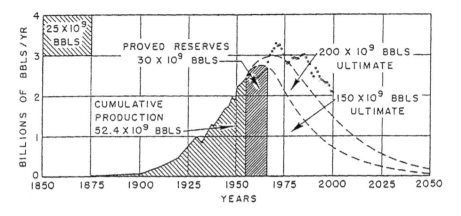

Figure 6.1 M. King Hubbert's 1956 graph of the Peak Oil problem.

is in for a major upheaval in every aspect of what we take to be modern life. So it's no trivial matter to understand why Hubbert got it right, and what implication that type of prediction has for the social arena today.

Hubbert's Peak Oil prediction shows in two different ways the crucial manner in which causation/inference enters into producing a useful prediction in the worlds of both natural and human affairs. From a purely geophysical point of view, Hubbert used known principles of how oil is formed, the type of geological regions where it's likely to be found, and other such geophysical properties, together with estimates (some would say "guesses") at consumption rates, known reserves, and the like, in order to forecast that US reserves would pass the halfway mark in the 1970s. For all practical purposes, he turned out to be right.

Notice the chain of implication here. We have

natural processes + guesswork $\longrightarrow$ **peak oil + guesswork** $\longrightarrow$ **social upheavals**

So an event at one level, suitably seasoned by inspiration and educated guesses, leads to predictions of events at another level. And that prediction in turn serves as the source for predictions of events at yet another level. Thus, if the forecast many make today that global oil reserves peaked in 2000 is as correct as Hubbert's prediction for the USA, such predictions have implications for how life will look in the second half of this century. Without viable alternative energies waiting to take over as the availability of oil declines, we can expect a very different world: A major falloff in long-distance travel, the outbreak of international warfare to secure what resources remain, the likely appearance of a worldwide pandemic, the disappearance of suburbia, the withering away of globalization, and the vanishing of the consumer economy are but a few of the consequences of turning off the cheap-energy spigot.

Of course, these forecasts themselves are based upon their own assumptions as to the likelihood of discoveries of major new oil reserves, the development of currently embryonic energy technologies, and beliefs about how people will react to sky-high energy prices. These speculations run parallel to Hubbert's estimates about oil reserves and consumption patterns. But in both cases events and conjectures about nature and human beings are combined to project other events some years into the future.

The point here is that forecasting the pattern of events involves a combination of educated guesswork (chance) and demonstrable laws of nature and facts (necessity). Let's look into this combination in a bit more detail and relate it to our overall hypothesis about social mood and its biasing effect on *human* events.

LIFE GOES ON

The bank robber Willy "The Actor" Sutton once replied to an interviewer's question about why he robbed banks by saying, "Because that's where the money is." While there's little doubt that there is indeed a lot of money in the vaults of banks (although somewhat less after 2007), if you're looking for who has the money don't go to the banks, go to the life insurance companies. Among the reasons these insurance companies have such bloated coffers is that they do a damned good job of prediction. In particular, they are able to predict death rates for groups of individuals with very high accuracy. Figure 6.2 illustrates this power showing an error rate of only 0.3 deaths per thousand people for an average rate of ten deaths in a thousand people. But the lower part of this diagram, which is simply a magnified version of the upper part, shows that forecasters have almost no skill at all in predicting *changes* of the death rates themselves. The forecasts consistently miss the turning points in death rates, especially after about 1970. This issue of turning points is crucial as the real payoff in prediction is knowing when trends are about to change, rather than simply knowing the trend itself.

So prediction of death rates is a mixed bag: The overall rate is easy to predict because it's pretty much constant at a low resolution. But the

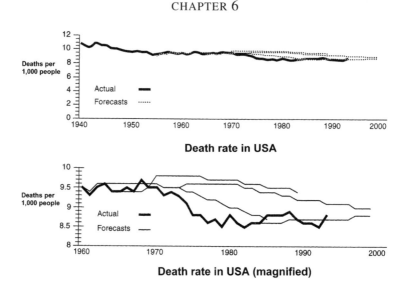

Figure 6.2 Death rates in the United States and missed turning points.

turning points are not. What about birth rates? Will they yield to the demographers statistical wizardry? Regrettably, the story is essentially the same as with death rates, especially when it comes to accurately forecasting the all-important points at which birth rates shift from "the same as last year" to something very different.

To find true success in predicting a social phenomena we have to look far and wide—and at the largest and smallest possible scales. For instance, it has proven to be straightforward to predict overall world population. Forecasts are reliable decades in advance because there is no uncertainy as to migration (since, among other reasons, no one is yet leaving for outer space, physically anyway). But when it comes to predicting the population change in individual countries (a finer-grained phenomenon), things start to go awry. And this degradation in accuracy gets even worse when the magnification knob is turned up to look at population changes in individual metropolitan areas. So where *do* we find any measure of success in looking into the future using traditional mathematical and statistical tools? The short answer is, "In the natural sciences." Moving to another hemisphere and another continent, here is an example of accurate, useful forecasting in action.

The graph in Figure 6.3 shows the success of 48-hour rainfall forecasts in Australia for the years 1970–1986. Here the scale of evaluation is something called the normalized S1 Score. This value is the departure of the forecast

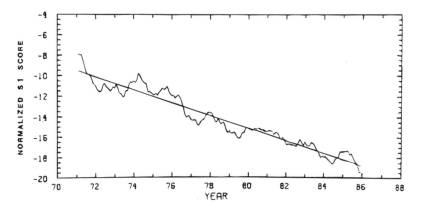

Figure 6.3   Rainfall forecasts in Australia, 1970–1986.

from the persistence forecast, which is simply the "no-brainer" prediction
that tomorrow's rainfall will be the same as today's. Large negative values represent large improvements over a persistence forecast. Notice how
improved observations and vastly better computing capabilities led to dramatically improved forecasts over this 16-year period. In general nowadays,
useful weather forecasts can be made for a period of up to about 7–10 days,
after which instabilities in the model, errors in the observations, and missing
pieces in the equations governing atmospheric behavior combine to degrade
the forecasts.

Atmospheric predictions over short timescales of a few hours to a few
days fall into the category of *weather*. Longer timescales, such as years
and decades and/or larger spatial areas like whole continents, lead to questions of *climate*. As we saw above in relation to population forecasts, it
is generally the case that the coarser the resolution, the more accurate the
forecasts. So we might expect to see even better forecasts of climate than
of weather. At first sight this may strike the reader as just the wrong way
around in view of the very heated—and sometimes acrimonious—debate
nowadays over quintessentially climatic phenomena like global warming
and the greenhouse effect. But, in fact, there are good examples to show
that climatic forecasting is at least as reliable as weather. Here's one.

Raymond Wheeler was a psychologist at the University of Kansas back
in the 1920s, who was investigating the effects of climatic change on human
behavior. In an attack on the behavioral psychology theories in vogue at
the time, Wheeler claimed that human behavior was not built up piece by

167

piece from elementary primitives, but was an integrated pattern of responses to the individual's environment. This view forced Wheeler to consider whether fluctuations in the environment might play a determining role in the way people behave, a notion that caused him to look at the role of climate variation in human affairs. To his great surprise, Wheeler discovered strong correlations between weather cycles and historical factors such as types of governments, wars, cultural styles, and human achievements. More specifically, he discovered a 100-year cycle, divided into four almost equal parts, indicating that humankind has behaved differently during periods of warm-wet, warm-dry, cold-wet, and cold-dry weather.

The Drought Clock in Figure 6.4 illustrates Wheeler's theory, correlating weather with the major wars of history. In the diagram, C stands for

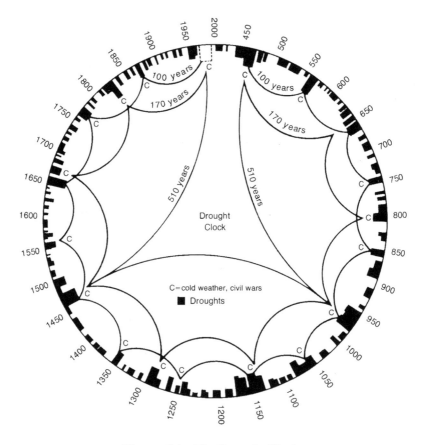

Figure 6.4 The Drought Clock.

cold periods. As the Clock shows, most international wars are waged during warm periods, while civil wars take place during cold-dry times. The Clock also shows that a major cold-dry period with political anarchy occurs about every 510 years. Perhaps ominously, the Drought Clock also indicates that we are now entering a major cold-dry era—a time of great civil unrest, lower agricultural productivity, and social revolution. If newspaper headlines over the past few years are any indicator, Wheeler's theory certainly should not be dismissed out of hand.

These examples, especially Wheeler's theory of climate in the affairs of humans, underscore the claim made earlier that events at one level can serve as causes for events at another. We see from these examples another pivotal feature of predictions. In general, if you want to be able to consistently make useful predictions, then you must sacrifice details in the forecast. This is rather trivially the case for weather forecasts, since it's extremely difficult to predict temperature and humidity with high accuracy for every part of your backyard. It's a lot easier to get it right on a continuing basis if you broaden your horizons a bit and consider your entire town, since in that case the weather patterns are averaged out over a lot of backyards.

This tradeoff is even easier to see in social and technological forecasts. If you state that something rather specific will take place, say colonization of the Moon by 2020, then you almost surely will have a high failure rate. But these are just the types of forecasts that attract attention because they have high information content; they tell you something surprising that you'd truly like to know about. On the other hand, if you reduce the information content in your predictions to assertions like "It's possible that humans will exploit resources in space sometime in the next millennia," then you will have a pretty high hit rate. But such a prediction is far too vague and the time frame much too long for the forecast to contain any really useful information.

Wheeler's Drought Clock also shows in literally graphic form one of the other major blunders that forecasters often run afoul of, namely, the confusion between a cause and a correlation. For instance, one of the central messages of Wheeler's theory of historical cycles is that there is a high correlation between periods of dry-cold climatic conditions and periods of social anarchy. Fine. There may also be a high correlation between short skirts being fashionable today in Italy and wheat prices shooting off the scale in Argentina. But can one create a believable chain of causation by which Argentine wheat *causes* fashion designers in Milan to favor short skirts, or vice-versa? Difficult, very difficult. But it may be rather less difficult to create such a causal chain between cold-dry climatic conditions over decades and collective group behavior leading to the destruction of old political and governmental structures. In fact, causal chains of this very type are what Wheeler argues in his book cited in the References for this chapter.

The point here is that without such a plausible chain of connection between the putative cause and its effect, you have at best only a correlation. The evidence presented in this book already supports strongly the case for correlation; the research program for our Central Hypothesis presented in Appendix B is aimed at "upgrading" these correlations to a strong biasing factor, if not an actual cause, since science is for the most part focuses on repeatable patterns that can be *explained* through causal relations linking things like mood and subsequent collective events. This point is crucial for any effort to turn socionomics into an actual science of collective human behavior. Let's look just a bit deeper into the matter by way of a very illustrative example.

CORRELATIONS, CAUSES, AND CHANCE

On November 29, 1978 an earthquake measuring a whopping 7.8 on the Richter scale rocked the Pacific coastal state of Oaxaca in Mexico. While this quake was an unmitigated disaster for the people of the region, it served as a major milestone in the centuries-long struggle by seismologists to understand the nature of earthquakes and to predict accurately the time and place of these geophysical rumblings. To understand why, let's quickly review the prevailing reasons for how earthquakes happen.

According to the theory of plate tectonics, the world's land masses can all be thought of as gigantic slabs floating about on top of a denser, semi-molten material in the Earth's mantle. As these slabs move about, they collide with one another in various ways. Head-on collisions form some of the world's great mountain ranges, like the Himalayas. Oblique collisions give rise to more ominous effects—they generate earthquake zones. Generally, in these oblique encounters the slabs collide and stick together. As time goes by, this bond is put under increasing stress as the two slabs try to move apart in different directions. Eventually the bond fractures under the tension, resulting in a rapid, jerky type of shift of one slab relative to the other. In other words, an earthquake. This picture indicates rather clearly why the magnitude of the resulting earthquake is directly dependent upon the strength of the bond between the slabs and the slabs' relative speed and direction of movement.

A fairly recent discovery in the earthquake-modeling business is that in the region where the two slabs are joined, even small stress-relieving tremors may cease shortly before a major quake. This quiescent period, termed the *alpha phase*, may persist for a number of years. Then a *beta phase* begins in which a number of small shocks occur over a period of a few days to a few months, culminating in a major quake. In 1977 a group headed by Gary Latham at the University of Texas Geophysical Laboratory programmed a computer to look through the world's seismological data, searching for likely-looking seismic "gaps" of this sort as indicators of major earthquakes in the making. The computer discovered one such alpha-phase gap for a region about 200 miles long on the coast of the Mexican state of Oaxaca.

The Texas group published its findings along with a very precise prediction of where it expected the quake's epicenter to occur: Between 16° and 17° degrees north latitude and between 96° and 97° degrees west longitude, an area on the coast smack dab in the middle of the Oaxaca location. The seismologists further predicted that the magnitude of the quake would register between 7.25 and 7.75 on the Richter scale—a major jolt. The only thing they didn't predict was the exact timing of the event, although they did note that the quiescent phase had already existed "substantially longer" than similar time periods in the past that had preceded large quakes.

At this stage the ugly head of politics began to intrude into the affairs of science. The Mexican press got wind of the earthquake prediction and splashed it across the front pages of newspapers in the capital, Mexico City. The result was panic in Oaxaca, as a number of residents sold their homes at bargain-basement prices and fled the area. Many local politicians denounced the prediction as irresponsible, a sentiment reflected in the words of experimental seismologist Karen McNally when she remarked, "Predicting an earthquake in somebody else's country ... is something you do with great trepidation." Ironically, McNally herself was to play a central role in carving out the scientific triumphs later salvaged from the rubble of Oaxaca.

Operating on the basis that there's no such thing as too much data when it comes to testing models of earthquake prediction, McNally and her co-workers set up a number of monitoring stations in the suspect region, including one right on top of the spot where the Texas group claimed the quake would be centered. Its instruments in place by early November 1978, the McNally group started seeing increased seismic activity around November 15. The activity then decreased until November 28, when there was another round of shocks. The next day there was a major quake located exactly in the center of the area where McNally's group had placed its array of monitoring stations—the strongest possible vindication of the concepts and methods employed by Latham and the Texas group.

While this story only scratches the surface of the human, political, and scientific drama associated with the 1978 Oaxaca quake, it serves admirably as a case study of how the scientific world understands the terms *explanation* and *prediction*. In brief, the theory of plate tectonics was used to *explain*, i.e., provide a *cause* as to how the earthquake would occur and where it would strike. An entirely different theory involving the alpha and beta phases of activity was employed to *predict* the timing of the quake. Since both the explanation and the prediction were based upon scientific theories and their concomitant laws, a natural starting place for considering scientific explanation, in general, is with an investigation of the lawful relations between events and the theories used to explain such lawful relations.

As we noted earlier, in everyday conversation we think of an explanation of some event as the answer to a question that begins with Why. Why is my wife yelling at me for not taking out the garbage? The answer to such questions is a cause for the observed event like, Because the remnants of last night's trout amandine are smelling up the kitchen. Or, Why haven't I heard from my broker recently? Because my portfolio has declined 10% over the past week on the strength of his hot tips. Simple cause and effect.

Perhaps the first such causal scheme was that advocated by Aristotle in his *Physics*, where he identified four collectively exhaustive and mutually exclusive causal categories: *material, efficient, formal*, and *final* causation. Each of these categories supplies the answer to a "Why?" question, and together they were supposed to give a complete explanation for why things are as we observe them. To use one of Aristotle's own favorite examples, we might ask "Why is my house the way it is?" The Aristotelian answers are that the house is as it is because of the materials from which it is constructed (material cause), because of the labor that the workmen put into building it (efficient cause), because of the architectural plans describing the structure of the house (formal cause), and because of my desire to have a comfortable shelter from the elements (final cause). In this setup we see the essence of the commonsense view of explanation: causal connection.

The key concept in science is that of a law, and all scientific explanations as to the cause of things invoke scientific laws to legitimize their assertions in one way or another. But not all laws are causal, and it has been forcefully argued that our preoccupation with causal laws is deeply tied up with the subconscious tendency of us humans to re-create the universe in our own image. Thus, since *we* feel psychologically comfortable with cause-and-effect relationships, we try to impose that pattern on nature as well. We'll come back to this point in a moment. For now let's take a short detour into the labyrinthine world of causality to look at a couple of the more notable landmarks.

What is it that typifies causality? Given two events A and B, how can we say that one of them acts as the "cause" of the other? Again speaking in everyday terms, we would probably say that A causes B if the occurrence of A at one moment *necessarily* brings about the occurrence of B at a *later* time. Note here how the idea of law has been brought in to link the two events: A today *requires* B tomorrow. Employing the terminology of logic, we would express this lawful relationship by saying that A *implies* B, showing that causality in the physical world is intimately tied up with the idea of deduction and inference in the austere world of logical abstraction.

As the distinction is absolutely central to our ability to predict natural and human events with confidence, let us once again compare the idea of causality with that of *correlation*. Suppose we have two events A and B such that whenever A takes place, B is always observed to follow. To take a familiar example, in one study it was noted that there is a high positive correlation between the number of storks seen nesting in English villages and the number of children born in these same villages. Such a correlation is rather similar to what we would term an empirical law or relationship. But in today's high-tech world, it's difficult to see any more of a causal connection between storks and infant births than between black cats and bad luck. There is just no such causal relationship, at all. What's lacking here is a *theory* that yields one or more laws, which taken together would provide a logical chain leading from the appearance of storks in the villages to the appearance of babies.

We have now seen that patterns in the physical world like the fluctuation of climatic conditions can often serve as candidate causes for major types of events in human affairs, such as wars, political upheavals, and even the destruction of nations. But what about human events of a similar sort that are not caused by natural forces but by factors inherent to the human system itself? Do these events also occur in some repetitive fashion that we might term a *pattern* or *cycle*? And if so, can we identify causes in the social domain that lead to these patterns? We've taken up the second question as we made our way through this book. Now we want to summarize what we've learned from the spectrum of political, social, and historical examples we've examined at in order to see how to forecast, or at least anticipate, such events in the days, months, and years to come.

Now it's time to give some relief to the reader impatient to leave the philosophy of science and return to social mood and its utility in forecasting collective human events.

THE ANTICIPATION MACHINE

Speaking one time about making predictions, the legendary Danish quantum physicist Niels Bohr said, "It's very difficult to predict—especially about the future." This was from a man who had a horseshoe hanging over the door to his house because, as he told a friend who wondered whether Bohr really believed such a charm brings luck, "They say it works even if you don't believe in it." The same might be said of our social-mood-based ideas to explain the development of trends, and even in some cases specific events. It seems to work—even if you don't believe in it!

Looking at indicators of social mood in the early 1980s, Robert Prechter saw historical levels of pessimism and a pattern in the stock market, which led him to anticipate a shift to a more positive mood with consequences that would specifically include, among other things, an economic boom and no international war for at least 10 years. Examining indicators of social mood in the 1990s and early 2000s, he noted the extreme levels and duration of optimism. Under these circumstances one might expect a shift to a more negative mood, with all its attendant consequences. Timing and extent would depend upon the specifics of price patterns and their degree (timescale), along with the probabilistic nature of conclusions to be drawn from them. And every day brings new information. The socionomist wishing to anticipate social events needs to be alert to new information and flexible enough to accommodate it. (Un)fortunately, doing so in a consistent manner is a daunting task.

In this section readers will have a chance to put to the test their understanding of our Central Hypothesis. Here's the plan. I will present charts of the DJIA on several different timescales to be used as a measure of the social mood. I'll also give a table of mood polarities suggesting how the events that follow from these positive or negative social moods might be categorized in everyday language. Following these background items, I'll set out a list of basic questions, each addressing a specific type of social behavior or action, together with a snippet or two of information taken from the media that would appear to relate to the question at hand. The reader's job is then to use this raw material in order to propose an answer to the question. In short, you're to give an answer as to what you believe will be the most likely way things will turn out.

Just as with weather forecasting, there is no one best answer but only a spectrum of possible answers, each with its own likelihood. So what we're doing here is not a "mid-term exam" in the usual academic sense, for which the questions actually do have rather specific answers (or at least answers the professor is looking for). Rather, this is a test of understanding the methods and tools of socionomics, as well as a test of how well you have assimilated socionomic thinking into your mindset about the world as it actually is and how it might be in the future.

As we are now moving into the much murkier terrain of looking into the future rather than raking over the coals of the past, let me again emphasize that socionomics is *not* really about prediction, just as science itself is not fundmentally about prediction. Both are much more about answering a huge question: "Why do we see what we do and not see something else?" Notice that this is a "Why?" question. That means we're asking for an explanation, not a forecast. This having been said, sometimes a good explanation of an observation generates at the same time an ability to at least partially foresee what is more or less likely to turn up next. What follows should be taken in this spirit.

Our leitmotiv will be to recognize that the progression from a positive social mood to a negative one is a movement from one polarity to another. We can label these polarities with everyday English words appropriate for the particular type of social trend or event we are focusing upon. So, for instance, if our concern were with the future of the automotive industry, a firm like Ford or BMW might be thought of as *adventurous* or *outward-looking* in times of rising social mood as they try to expand their markets to other products and/or territories. On the other hand, as the mood turns negative we might see a situation in which these firms become *protectionist*,

Table 6.1 Polarities of positive (+) and negative (−) social moods

Positive mood (+)	Negative mood (−)
Unifying	Divisive
Liberating	Restricting
Adventurous	Protectionist
Togetherness	Separation
Supportive	Opposing
Open	Closed
Happy	Sad
Hard-working	Lazy
Manic	Depressive
Tolerant	Bigoted/xenophobic

seeking to preserve whatever market they already have against real or perceived "poachers." So the two terms—adventurous/protectionist—represent polar opposites associated with strong positive or negative social moods. Table 6.1 gives several more such polarities.

I'm sure that by now the reader can fill-in many more entries in this table. But the general idea is clear. As we move from positive to negative mood, we move from a state of optimism and looking forward to a brighter future to pessimism and a future we fear. This psychological disposition in a population has deep implications for the relative likelihood of societal events of every sort taking place. The implications we draw from this, however, should all be taken with a shakerful of salt for the reasons we've already discussed. The more specific the type of event we're focusing upon, the less likely it is to occur as forecast, at least in fine-grained detail. But what is likely is that an event from the same "family" of events is more likely to occur than something from a very different family.

This labeling of the character of an event is a bit like trying to foresee the precise details of how a child will look as an adult by knowing the genetic makeup of its parents. If the parents are tall, you would expect the child to be taller than average too. But the child's height as a fully-grown adult is difficult to predict with any degree of precision. Similar is the case for other traits such as hair color, intelligence, and physical build. So as the reader attempts to answer the questions posed below, it's good to keep this analogy in mind.

We also noted before that our Central Hypothesis involves social events unfolding on very different timescales. So the assessment of whether the social mood is positive or negative depends on what timescale we're talking about for the development of a particular type of event. It can easily be the case that the mood lies on the positive side of the foregoing chart for, say, a short-term event like a developing trend in tastes in popular films, while being very negative indeed for longer-term events like peace, or at least the cessation of hostilities finally coming to the Middle East.

Some may think that this same argument could be reversed, in the sense that for any event there is some timescale for which the mood is positive and uplifting for that event, and another timescale for which it is exactly the opposite. If we were talking about *specific* events, then there might be a germ of truth to this argument. For instance, in political election campaigns there is usually a timescale on which the fortunes of a candidate are rising and another on which they are falling. But our concern is not with such very definite things as a particular candidate's ups and downs, since they are basically random flucutations superimposed on a stronger and steadier underlying pattern. Our interests are in the signal (pattern), not the noise.

At the risk of undue repetition, note again that our focus is on collective social behaviors, not specific events like assassinations or lottery winners involving individuals. And those behaviors unfold on a scale characteristic of the particular type of behavior. Wars do not break out in hours or days, they take weeks and months or even years to actually unfold. Similarly, a change in popular taste in music does not reveal itself over centuries; it is a phenomenon whose natural timescale is measured in weeks to at most a few months. So in evaluating the questions that follow, you should first ask yourself, "What is the natural timescale for the event in question?" At that point you can go to the chart of the DJIA or another financial index perhaps more relevant to the event in question to use as your sociometer, examining it on the relevant timescale for the event in order to evaluate whether the mood is rising or falling. This information, together with the polarities of Table 6.1, as well as any other supporting information you have, such as media reports, personal observations, and the like, constitutes the raw material from which to formulate your answer to the question.

To illustrate this point of timing, as well as to set the background for testing projections of different types of events in the coming years and decades, here is a series of graphics showing the DJIA on radically different timescales. The first is the long-term view shown in Figure 6.5, in which the "ticks" on the timeline are measured in half generations of 15 years each.

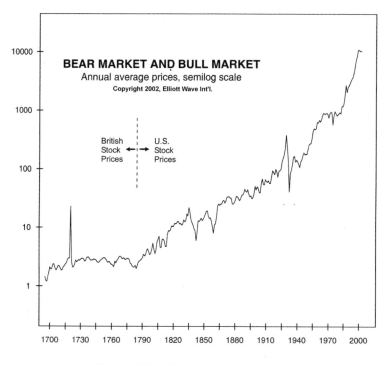

Figure 6.5 The DJIA, 1720–2000.

(Note: The data on this chart are for British prices for the period ending around 1784. After that the data are from the US markets.)

The picture shown in Figure 6.6 is for a shorter timescale measured in steps of decades, not generations. Zooming-in on Figure 6.6, the picture in Figure 6.7 shows the yearly fluctuations of the previous decade-scale chart. It's important to observe that even though the longer timescale movement is trending upward, indicating a generally positive social mood on a scale of decades, within that same period the shorter timescale chart shows many periods of a year or two when the trend is actually down, indicating a "mini" negative social mood. And if we were to look at even shorter timescale price movements we'd see these positive and negative moods themselves split into similar ups and downs. This, by the way, is the very essence of the fractal nature of financial wave patterns.

Now back to forecasting. Here is the background against which all the examples to follow are projected:

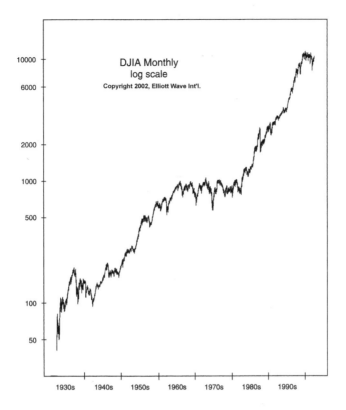

Figure 6.6 The DJIA on a decades scale, 1930–2000.

Fundamental Working Hypothesis: *There will be an accelerating shift from the positive to the negative polarities in* Table 6.1.

If it is indeed the case that the next few decades will see an accelerating movement from a positive to a negative social mood, that movement has implications for every area of human life. Some of those implications, along with recent observations supporting them, are part of what the reader is now going to be called upon to judge in addressing the set of test questions. A reader who disagrees with our Fundamental Working Hypothesis and feels that the social mood will rise rather than fall—or perhaps remain unchanged—can simply reverse the implications of the declining mood. That's the power of the method. Get the social mood and the timescale right, and everything else will follow pretty much automatically—at least in a probabilistic sense. Finally, we turn now to the question-and-answer part of this section.

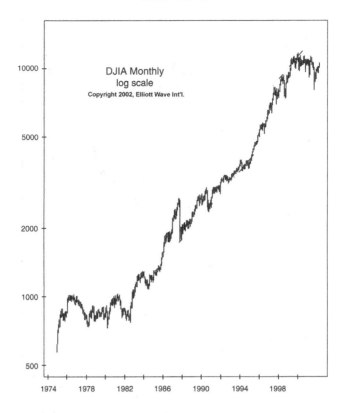

Figure 6.7 The DJIA on a yearly scale, 1974–2000.

For each test question, first there is a statement of the question, followed by an item or two of information taken from the media that seems to be relevant to answering the question. Note that here I'm using media accounts from about 5 years ago, so you have a chance to see how well they've played out in the intervening period of time. I'll give some arguments for why things have happened the way they have in the next section. For now, let it suffice for me to state (without proof!) that the headlines (read: events) today reflect the mood yesterday. So when you read about something in the newspaper, the social mood giving rise to that "something" has already been factored-in. In short, they are *lagging* indicators.

Just so you will get into the habit of evaluating your arguments in the face of perhaps strongly-held opinions to the contrary, I'll also indulge in a bit of editorializing about the media report(s). These mini "editorials" should leave no doubt as to where I stand on the question at hand. But that's no reason for you to follow my line. Anticipation/forecasting is a game anyone can play. And how well you do depends on how well you employ the Central Hypothesis with the data available.

Now here are the questions.

Will the price of oil drop dramatically over the next 5 years?

Figure 6.8 shows a very long-term chart of crude oil prices since the discovery of oil in 1859 until 2009. This is one piece of evidence to consider in evaluating the question. Here are a few media statements to also throw into the hopper.

Media Item: The period 1982–2007 will surely go down in economic history as what Prechter and Kendall call the Great Asset Mania. (The sheer size of their more than 700-page coffee table book, *The Mania Chronicles*, attests to the Mania's all-time status.) During the market selloff in 2001–2002, there was a drop in prices. But from late 2002 the Federal Reserve has made unbelievable amounts of easy credit available in an attempt to recapture the "good old days." In the real-estate sector one might even term this a kind of *levitation*, as credit has resulted in prices soaring literally off the chart. But the reflation effect of massive credit is coming to an end, with

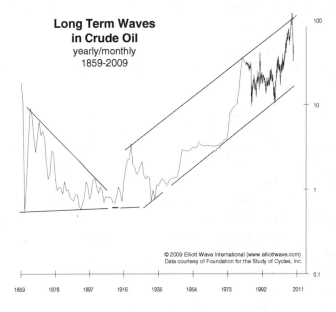

Figure 6.8 Crude oil prices, 1859–2009.

the economy and markets, and concomitantly the social mood, on their way into a *deflationary* depression that will last for many years.

Evidence in support of this media claim is graphically evident in Figure 6.9. Here we see the price movement of seven very different markets from stocks to bonds to commodities for the period late 2001 to November 2008. In every case, the price trends reversed completely from an upward flight earlier in the decade. Junk bonds started the move south relative to Treasuries in 2005, while stocks peaked in 2007 and commodities in 2008. The total downturn in all these financial instruments is something that no one alive today has ever seen.

Most would argue that to see junk bonds and commodities both falling together, along with oil and transportation stocks, is simply not possible. The conventional wisdom has it that these instruments move counter to each other. So not all of them can fall (or rise) together. But the chart is very clear that they can certainly all rise together—just pump enough easy credit into the system and get someone, somewhere (mostly Asian central banks), to pay the bill and you can literally suspend the normal rules. But anything that goes up must come down. And when "anything" becomes "everything," that is precisely what happens. It even has a name: Deflation!

Is oil likely to be an exception? Or will it fall along with precious metals, real-estate prices, bonds, and, of course, stocks, Peak Oil not withstanding?

Next question, please.

Will openness and transparency yield to increasing secrecy in all spheres of life, especially in government and business, over the next decade?

Media Item: A headline in the July 4, 2005 issue of the *International Herald Tribune* stated, "Official Secrecy Reaches Historic High in the US" The subhead is still more revealing: "Even CIA Budget from 1950s is Off-Limits." The thrust of this piece is that the government, driven by terrorism, has resorted to an historically high level of secrecy. An astonishing statistic mentioned in the article is that 125 documents are being classified *every single minute*. A total of 15.6 million such documents got the censor's stamp in 2004 alone, nearly double the number classified in 2001.

It's truly amazing to read some of the kinds of things the US government thinks are worthy of holding back from the public eye: the CIA's budget from the Cold War era, the Defense Intelligence Agency's report that former Chilean dictator Augusto Pinochet was interested in fencing, boxing, and horseback riding, and the Justice Department's adamant insistence that a four-line quotation from a *published* Supreme Court decision be deleted

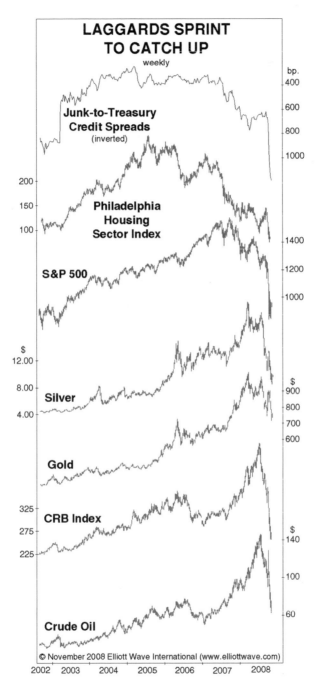

Figure 6.9 All the same markets for seven commodities.

183

from a document, to name just three examples of many. Some commentators lay the blame for this onto hawks from the Bush administration, like Vice-President Dick Cheney and former Attorney General John Ashcroft.

As a budding forecaster, do you think this penchant for keeping things hidden is a direct consequence of declining optimism in the populace, at large? Is it still one more implication of an increasingly negative social mood that will last for generations? For just a bit more perspective on this sub-question, here are some bonus media items on the topic.

Media Item: An editorial in the July 8, 2005 issue of the *International Herald Tribune* denounced the jailing of reporter Judith Miller of *The New York Times* for refusal to divulge a source. In their commentary, the editors observed that Ms. Miller was surrendering her liberty in defense of a greater liberty, that granted to a free press by the Constitution to work in the public interest without fear of regulation or retaliation from their government. So much for freedom of the press, at least in today's—and tomorrow's—America.

Media Item: At a news conference at the end of 2005, President George W. Bush defended the government's right to spy upon individuals as being "critical to saving American lives." In reaction to this unapologetic declaration, one Democrat said Bush was acting more like a king than a democratically elected leader. Undeterred by the criticism leveled against his administration for the eavesdropping on US citizens without benefit of court-approved warrants, Bush went on to make it clear that he had no intention of halting the monitoring activities and even made the forceful claim that revealing the program to the public by the news media endangered Americans.

Media Item: Writing in the journal *Terrorism and Political Violence*, John Mueller notes that in any given year the number of people worldwide who die from terrorism is not much more than the number who drown in bathtubs in the USA. In a similar vein, columnist William Pfaff noted that since the late 1960s, the number of Americans killed by international terrorists—including 9/11—is about the same as the number killed by lightning or by severe allergic reactions to peanuts! Does this sound like World War III? Or as the Pentagon put it, a war against "The Universal Adversary"? To some it sounds a lot more like a marginal adversary that has been conflated by the US government with the social upheaval in the Muslim world, in order to institute changes in American society that are vastly more dangerous to the American way of life than Osama bin Laden will ever be.

So how about it, dear reader? Does this sound like a happy, open environment? Will the next decade be Orwellian, Panglossian, or perhaps neither? Here's another teaser along the same general lines.

Are family members likley to be forced to move in with each other with increasing frequency during the coming decade, as homelessness increases due to skyrocketing prices for affordable housing?

Media Item: A major article in the June 26, 2005 issue of the *International Herald Tribune* notes the problems parents can (and do) face when their grown children start moving back *into* the family nest rather than out of it. In the United States alone, nearly 16 million families had at least one child over the age of 18 living at home in 2003, a 7% jump in a single decade. While figures are not available for other countries, anecdotal evidence suggests similar increases in Britain, France, and Spain. For instance, the Royal Bank of Scotland surveyed 1,200 university graduates in the UK and found that nearly 60% are still being supported by their parents—3 years after graduation! And this is nothing compared with the sad plight in Italy, where over 80% of single men between the ages of 18 and 30 are still living at home. The financial and emotional burden for parents of these "boomerang" children is likely to be a continuing theme in family life during the coming decades.

Media Item: In spring 2005, the Los Angeles police department reported that a mob of 3,000 people showed up to claim one of 56 affordable housing units. A captain on the force said, "You had some very desperate people who had a mob mentality. It was as if people were trying to get the last piece of bread." This is all reminiscent of a fear of rising prices in March 1837, when protesters in New York City marched with signs demanding "Bread, Meat, Rent, Fuel! Their Prices Must Come Down." Readers might ponder the fact that the riots in 1837 were followed by the Panic of 1837, the opening salvo in what became one of the greatest deflationary periods in American history.

So should parents start adding-on another room for the return of their prodigal children in the coming years? Or is affordable housing going to make a comeback?

Are old-time, fundamentalist religions going to become fashionable again, with believers in different creeds becoming increasingly intolerant of each other?

Media Item: Citing a law forbidding "outrages against religion," an Italian judge in Bergamo ordered the noted journalist Oriana Fallaci to stand trial over a book in which she made provocative assertions about Islam. In an earlier volume, *The Rage and the Pride* written in the wake of the 9/11 attacks, Ms. Fallaci questions the ability of Islam and western society to be reconciled and took Europeans to task for failing to defend their culture. The book in question, *The Force of Reason*, revisits much of the same ground. In this work she states, "To be under the illusion that there is a good Islam and a bad Islam, or not to understand that Islam is only one or the other is contrary to reason." Apparently this was enough to lead the judge to order charges brought against her in response to a suit filed by a Muslim activist.

What can one say about such a case? In times of upbeat social mood it's doubtful that anyone would have questioned an author's right to make unpopular or even intemperate statements. But, alas, these do not appear to be such times. Are they likely to be in the foreseeable future?

Media Item: In mid-June 2005 the Associated Press reported that the superintendent of the US Air Force Academy acknowledged that his campus is permeated with evangelical proselytizing, and that it will take years to rid the academy of religious intolerance. The story went on to note that evangelical Christians wield so much influence at the school that anti-Semitism and other forms of intolerance have become pervasive. In the locker room of the football team, a banner stated, "I am a Christian first and last . . . I am a member of team Jesus Christ."

Media Item: Pat Robertson is founder of the Christian Coalition of America and is a broadcaster known for his controversial public statements on the Christian Broadcast radio network. At one time he even declared his candidacy for the US presidency. As an example of his bombastic style, in October 2003 he called for the US State Department to be blown up with a nuclear device, and said that feminism encourages women to "kill their children, practice witchcraft, destroy capitalism and become lesbians." As if this isn't enough, in August 2005 Robertson urged American agents to assassinate President Hugo Chavez of Venezuela, calling him a terrorist. In response, Chavez's Vice-President Jose Vicente Rangel stated that Robertson is "a man who seems to have quite a bit of influence in that [the US] country," and then added the comment that "religious fundamentalism is one of the great problems facing humanity in these times."

Does this sound like the type of polarized behavior characteristic of a negative social mood? Given such behavior by governmental and religious authorities no less, it gives one cause to again ponder the reality of the separation of church and state. So is the theme song of the coming decades likely to be "Give Me That Old-Time Religion"? Or will it be "Old-Time Rock and Roll"?

Will there be increasingly strident attacks on science and a resurgence of magical thinking?

Media Item: A piece posted on the web site of MSNBC in late 2008 was headlined "Business Booming for Psychics." The article went on to say that while thousands of people are losing their jobs, the profession of fortune telling is booming. In tough economic times, it seems that the usual skepticism that plagues the fortune-telling profession is cast aside as people become desperate to see what the cosmos has in store for their economic futures.

Media Item: During the past few years there has been a veritable explosion of books challenging the mainline neo-darwinian theory of evolution. Titles such as *Darwin on Trial, The Edge of Evolution: The Search for the Limits of Darwinism, Evolution: A Theory in Crisis, Not by Chance: Shattering the Modern Theory of Evolution*, and *Darwinian Fairytales: Selfish Genes, Errors of Heredity, and Other Fables of Evolution* give a fair sample of this prodigious outpouring of the godly influence in how humans came to be humans on planet Earth. Not even the masterful literary skills of Richard Dawkins seem to be able to stem this anti-Darwinian tide.

Media Item: To Roman Catholics, St. Joseph has long been considered as a saint that focuses special attention on home-related matters. And in today's real-estate market, selling a home at a price greater than the owner's mortgage seems to fall within his or her purview. At least Cari Luna of Brooklyn, New York—who's Jewish by heritage and Buddhist by religion—thought so. After she and her husband held five open houses, even serving home-baked cookies at one of them, in a futile effort to sell their home, she invested in a St. Joseph's "real-estate kit" online and buried the 3-in. statue that came with the kit in her backyard. The kit also came with a special prayer. Some kits even come with a statue of St. Jude, well-known as the patron saint of lost causes. According to purveyors of these kits, business is booming. So much so, in fact, that one of them had to switch to online dealings because the 25 calls a day he was receiving were simply too much to handle on a personal basis.

The picture emerging from these shifts from rationality to magic and mysticism is a pronounced wave of distrust, if not outright hatred, of reason and science in the mindset of the masses.

Are terrorists likely to attack in the USA and Europe again "soon"?

Media Item: In Joseph Conrad's book *The Secret Agent*, published nearly a century ago, one of the characters comments on a terrorist act, saying, "It must be purely destructive. It must be that and only that beyond the faintest suspicion of any other object. Madness alone is truly terrifying, inasmuch as you cannot placate it either by threats, persuasions or bribes."

Is this the goal behind the style of today's terrorism? Random bombings, large and small, ranging from 9/11 and the July 7, 2005 London Underground bombings mega-events to everyday suicide bombings in Jerusalem, Bali, Baghdad, and elsewhere? Or is the madness within ourselves? Are these blasts meant to show how little we can rely upon the thin veneer of civilization, and by extension, upon ourselves to protect our way of life? Or is the goal simply to destroy a civilization that the terrorists hate because they have no role within it?

In the eye-opening volume *Civilization and its Enemies*, philosopher Lee Harris makes the point that the classical "liberal principle of national sovereignty is in the process of dissolution." Harris argues that this sovereignty is the basis of the traditional concept of the nation-state, which is characterized by the monopoly of physical force under the control of a single central authority. In the time of such a world system, the rules governing conflict between societies took the form of a traditional Clausewitzian war carried out as the policy of a central command, rational in its design and purpose, to change the behavior of another nation-state. But no longer. At some point in the latter part of the twentieth century, this notion of a nation-state disappeared to be replaced by what Harris terms the *honorific* concept of the state. Now the state is no longer a political entity that can defend itself against all threats, internal and external. Instead it is an entity called into existence simply by formal recognition of the international community—in short, by a *social consensus*.

It is in exactly this sense that one can speak of the "state" of Palestine, for example. As Harris notes, this point is best illustrated by the offical explanation of the Palestine spokesman for the actions of suicide bombers and other acts of terrorism in Israel and elsewhere as "acts of war." Of course, under the old regime of what constitutes a state fighting a genuine war, the Palestinians would have been wiped out decades ago or forced to make some type of accommodation with Israel that was based on the respective resources of the two combatants. So a group like the Palestinians may demand a state and even been given one by the international community.

But that state will only exist through the liberal conscience and the grace of the Israeli people, argues Harris.

Looking at the wars of the first half of the twentieth century, probably the greatest force generating them was the unprecedented pace of industrialization that transformed working conditions and political structures in ways that were uncontrollable. Among other things, these were the forces that led to Communism and Fascism. We are now at a stage of social upheaval of a similar scope. Again the revolution is being driven by technology, specifically the technologies identified with America that facilitate rapid communication, transport, and commerce on a scale that would have seemed magical just a decade or two ago. As political pundit Roger Cohen remarks, this technology is "prizing open closed systems and rendering visible everything that was once remote or inaccessible. The unknown is merely mysterious, but what is seen may be envied or hated." Cohen goes on to argue that such technologies are irreversible developments that have generated resentments, and that fanaticism has its roots in such rapid change since fanaticism is rooted in an absolute certainty that consoles the mind and spirit when everything around it is in a state of flux.

It is truly one of the great ironies of history that terrorists make use of the very technology that they want to destroy to deliver their horrific messages. Gruesome scenes of beheadings of the "infidels" are broadcast on the Internet, cellphones are used to detonate explosives to seed fear in western populations, as the bombings on the London and Madrid transport systems illustrated, jumbo jet technology is employed to bring down the World Trade Center. These and numerous other terrorist acts serve to underscore the point that the terrorists wish to destroy the very means they employ to disseminate globally the medieval symbols of their message. And that message is simply to destroy all forces that threaten the narrow, backward-looking Islamic society that they regard themselves as nurturing and protecting.

Since the United States is the most visible symbol of the lifestyle and technology the terrorists feel most threatened by, the sooner the American people realize that they have an enemy who wants nothing other than to destroy them, the sooner it will be possible to see that World War III is not between traditional nation-states with all the taken-for-granted ground rules surrounding such conflicts. Rather, this is a radically new type of war, one being waged by Harris's honorific state against a traditional nation-state. Electrical engineers would recognize this situation as an "impedance mismatch," in which one system is being enjoined with another. But the two systems are operating on entirely different principles and rules of engagement. The West cannot prevail in such a war until it comes to terms with the rules of the game, and begins to engage the enemy on its own terms.

One can certainly argue that this one is likely to drag on for the next 20 years or so, allowing plenty of time for major escalations that seem almost surely to include many more attacks on sites made vulnerable by the liberal conscience of western societies. A very unhappy by-product of this situation is the subsidiary possibility that the US is very likely to threaten more countries and that many more countries than now will begin opposing US interests, public statements by the President Obama notwith-standing. As Confucius' well-known curse expressed it, "May you live in interesting times."

In these examples we've laid heavy emphasis on news reports as qualitative harbingers of the ever-shifting social mood to go side by side with the quantitative measure provided by the financial market indices. Since these factors complement each other, let's have a closer look at the news reports side of the street to see how to use headlines as well as markets, which we've temporarily put on the sidelines in this discussion, to get a glimpse of the future.

THE END OF A TREND

Ralph N. Elliott pioneered a process for identifying the gyrations of financial price indices long before the term *market technician* gained common currency. Elliott's ideas are presented in some detail in Appendix A. Right now, what's of interest for us is a remark he made regarding the value of news: "At best, news is the tardy recognition of forces that have already been at work for some time and is startling only to those unaware of the trend." To phrase this statement in more modern parlance, news may often act as recognition of a change of one sort of trend or another, just the opposite of a sign of continuation of that trend.

So what we should be doing when we look at news reports is assessing the *psychology* of the event the news describes, as well as how journalists package that psychology in the news report. As pointed out by Drew Ross in an article on how to "unpack" the news, we should be asking questions like

Do either the event or its report express a feeling of optimism or pessimism? If so, is the report the result of an extreme in psychological mood? Why did the event occur at this particular time? Often the actual event being reported occurred some time before. So why are we reading about it only now?

To illustrate this connection between news and social mood, let's return to the art market, a topic we discussed briefly in Chapter 3. In August 2004, *The New York Times* ran an article titled "When Beauty and Value Are in the Beholder's Eyes," which addressed the fanaticism of art and wine collectors. The piece used what's termed the Mei/Moses Fine Art Index to track the changing value of art sold at auction since 1875.

The Mei/Moses Index shows that fine art has increased in value at about the same rate as the S&P 500 stock index for the past 50 years. One of the developers of the Index, Professor Michael Moses of New York University, stated that "For people who have a 20- to 30-year horizon, I have no problem talking about art as an investment," a statement that we would surely classify as optimistic about the future on his stated timescale. The majority of people who read this statement would most likely simply project this view of art-as-investment linearly into the future, and thus assume that art values will continue to rise. But a believer in social mood theory might well read this same statement and ask whether the trend in art prices is now topping out. A look at the chart of art prices versus the stock index in Figure 6.10 suggests this rise in prices is indeed a mature trend close to rolling over rather than one that can be expected to continue too much longer. Therefore, the story is one of a peak in positive mass psychology, with the news item heralding the end of a trend, not its continuation.

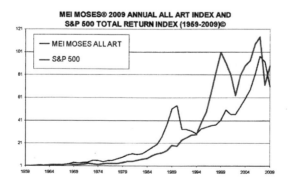

Figure 6.10   Art prices versus the S&P 500, 1959–2009
(Copyright 2009, artasanasset.com).

Ross identified a set of general guidelines for spotting the end of a trend in media headlines, giving each a colorful label to help the reader keep them in mind. Here is a somewhat edited version of his list.

Too Many Experts Spoil the Fun: When an article states that the major-ity of experts agree on a trend of anything, chances are the trend is rolling over. A good threshold level for the "majority" is 62%, the so-called Golden Mean.

What Everybody Knows: When the headlines all agree on a trend, the trend is topping.

The Last to Know: When news about a niche or specialty reaches the general press, the trend for that niche behavior is just about over. This is the so-called "Hot Tip Indicator," which says that when someone gives you a hot tip on something, be careful! Chances are the news you're receiving has already been discounted, and you're one of the last to know, not the first.

I Can See Clearly Now: When self-assured pundits and other talking heads claim that the trend will continue for the foreseeable future, it's time to get off the train. Such statements never seem to take into account that trends *do* change, and that a linear extrapolation of anything has to top out sometime. Why not now?

Reach the End to Begin Again: Statements that any social condition is permanent should be viewed as red flags. Examples? How about "The New Economy" for starters?

Well, this story about media accounts and the stock market as comple-mentary lagging and leading indicators could go on indefinitely. So let's lighten up a bit and return to our main thread of anticipating/forecasting what's likely over the months and years to come. A central ingredient in the process is to understand—and forecast—the social mood on the timescale appropriate for the event we want to forecast. As price movements on the stock market serve as our measure of this mood, this problem reduces to procedures for projecting those prices into the future—in other words, a reliable method for forecasting stock prices. This has been the Holy Grail of zillions of people for as long as stock markets have existed. And, inter-estingly enough, there are many procedures that seem to work—at least for awhile! To address this issue in our context of social mood forecasting, let's have a quick look at stock price prognostication.

HOW NOW, MR. DOW?

In late-summer 1967, Senator Thomas McIntyre of New Hampshire was participating in Senate Banking Committee hearings on the management of mutual funds. The thrust of McIntyre's inquiry was to question the extraordinarily high fees being charged by fund managers for the so-called investment services they performed for their clients. One of the witnesses before the committee was Nobel prize-winning economist Paul Samuelson, who stated categorically that there was no kind of publicly available information, price data or otherwise, that would enable fund managers to outperform the market.

Senator McIntyre, like most politicians, fancied himself a practical, no-nonsense type of guy. So while listening with great interest to Samuelson's theoretically oriented reasons for why a randomly-selected portfolio would yield results as good as or better than those of the funds, the senator did a little dart-throwing, picking stocks at random from the several thousand listed on the New York Stock Exchange, and then buying and selling them at random times. The outcome of his experiment showed that a hypothetical $10,000 investment 10 years earlier in the senator's dart-selected portfolio would have been worth a cool $25,300 by the time of the hearings—a rate of growth far in excess of even the most growth-oriented mutual fund. This was definitely bad news for the fund managers' attempts to justify their outrageously high fees on the grounds that they brought special expertise to the stock selection process. And, in fact, after this debacle the fund managers made no real effort to rebut Samuelson's testimony—although *The New York Times* reported that a couple of them did offer the senator a job!

The McIntyre incident vividly illustrates what has come to be termed the *efficient market hypothesis* (EMH). In everyday terms, the EMH is the claim that all information available is already reflected in the price of the stock. In the finance literature, the EMH comes in three different flavors each labeled as a certain type of random-walk theory (RWT). They are the

- *Weak RWT*: No technical analysis trading system based on price data alone can ever outperform the market. This weak version of the RWT is just the claim that you can't create information about today's prices from looking at what has happened in the past.

- *Semistrong RWT*: No trading scheme based on any publicly available information will be able to outperform the market. According to the semistrong RWT, not only are technical analysts useless, fundamentalists are too! But note that the semistrong theory applies only to *publicly available* information of the sort you see in *The Wall Street Journal* or that comes from scanning corporate earnings reports and listening to governmental agency pronouncements. The semistrong RWT constitutes what we earlier termed the efficient market hypothesis: All publicly available information has already been taken into account in setting the current price of the stock. But perhaps there's still hope for a winning portfolio selection strategy by employing so-called insider sources of information. For this case we have the strong RWT.

- *Strong RWT*: No trading scheme based upon *any* information sources whatsoever can outperform the market. Thus, the strong version says that no matter where you get your information, it will prove useless over the long haul in obtaining better-than-market-average investment results.

The perceptive reader will have already noticed that there's something inherently paradoxical about the EMH. On the one hand, the EMH says that it's useless to gather information; it will do you no good at all in the development of a trading strategy that will outperform the market. On the other hand, the EMH says that all available information has already been factored into the price of the stock. But how can this happen if no one gathers information? The fact is, it can't. Therefore, in order for the EMH to be valid there must be a sufficiently large number of traders who don't believe it! So it can only be true if you don't think it's true—a stock market version of the famous Liar's Paradox ("All Cretans are liars. I am a Cretan. I'm lying.").

Here we have the main pillars upon which academically-based attacks on the stock market are founded: the RWT and the EMF. Both of these ideas come with a host of built-in assumptions, like the EMH's inherent assumptions about the pricing mechanism and rationality on the part of investors.

If the EMH is even approximately true, then it should be impossible to make consistently better-than-average returns. But the empirical evidence clearly indicates otherwise: Wall Street is filled with market operators who do just that. How can we reconcile this fact with the EMH theory? According to the academic finance establishment, these above-average returns can be obtained only by accepting above-average risks. In short, the investor who achieves such returns is merely being compensated for a willingness to shoulder a level of risk higher than the market norm. Not so, say members of the vocal opposition. They say that there are so many stock market anomalies and just plain inconsistencies having nothing to do with risk premiums that it's necessary to look at the very idea of the EMH with a jaundiced eye. To adjudicate this debate, let's turn to a consideration of these matters and try to see if there might really be loopholes that we can slip through to actually construct an effective prediction/explanation scheme for beating the market. First we look at some forecasting rules for which timing is the central ingredient.

The traditional "chartist" methods against which the weak RWT argues involve plotting stock prices or indices as a time series of data. The chartist, or, using the more genteel term, *technical analyst*, then looks for characteristic patterns in these plots. Some typical patterns of the kind that such technical analysts look for are displayed in Figure 6.11. In these patterns,

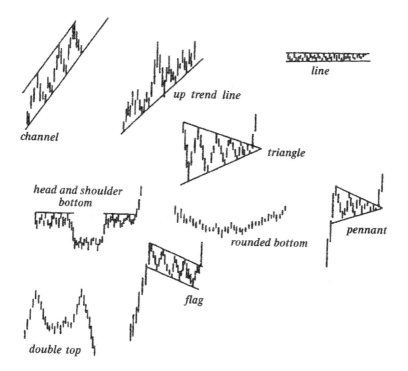

Figure 6.11   Typical chartist patterns.

the ends of the vertical bars represent the high and low levels of the price during the period covered by the bar, typically a daily or weekly time frame. While conventional technical analysts are convinced that the road to market riches is buried in the meanderings of such patterns, as Burton Malkiel, former dean of the Yale School of Organization and Management, notes, "I have never seen a successful technician yet."

To underscore his point, Malkiel tells the story of an experiment he conducted with some of his students. The experiment involved a hypothetical stock that initially sold for $50 per share. To construct a time series for the stock's price changes, Malkiel told the students to flip a fair coin. Heads would send the stock's closing price up half a point, while tails would send it down by the same amount. Following the experiment, Malkiel showed the resulting "chart" to a technician friend, who insisted on knowing the name of the company, frantically stating, "We've got to buy immediately. This pattern's a classic." But even though there's not a shred of evidence to support these traditional charting methods, there are a variety of other time-based patterns associated with market behavior that are not so easy to dismiss. Here are just a couple of them.

One of the most successful technical strategies for beating the market is the so-called *January Barometer*. This strategy is based upon a very simple principle: As January goes, so goes the market for the rest of the year. In other words, if the market is up for January, then it will be up for the year as well. Between 1950 and 1985, this technical indicator was right an astounding 31 of 36 times—nearly a 90% success rate.

In an attempt to make sense out of why January should be so important in presaging what the market will do in the coming year, Yale Hirsch looked at the market performance all the way back to the turn of the century. He discovered the following remarkable fact: Prior to 1934 the January Barometer worked only an anemic 19 times out of 33. But following 1933, all the way up to 1985 when the study ended, the success rate was a phenomenal 40 out of 50. Digging just a bit deeper, Hirsch offered a very plausible explanation for the dramatic change: the "lame-duck amendment" to the US Constitution enacted in 1933. Prior to this amendment, the Congress elected in November didn't actually take office until December of the following year, more than 1 year after its election. As a result of the amendment, legislators now take office only a few weeks after their election with the result that we now see many significant events affecting the economy squeezed into the month of January.

Another kind of technical trading scheme that has paid above-average dividends to its followers has been the so-called *Monthly Effect*. This indicator shows that the average return for stocks is positive only on days *immediately before and during* the first half of the calendar month, and is essentially zero for days during the last half of the month. As evidence, consider that during the period 1963–1981 *all* of the market's cumulative gain occurred just before and during the first halves of months, with the second halves contributing absolutely nothing to the advance. Subsequent studies have shown that this effect is real and cannot be attributed to biases like the January effect or the concentration of dividend payments in the first or last half of the month.

It turns out that there are many such technical indicators like the January Barometer and the Monthly Effect—all based on trading schemes involving certain days of the week, parts of the year, political changes, and the like—each of which seems to outperform the market averages. With this kind of empirical evidence from the technicians, it's difficult to avoid the conclusion that there's something dramatically wrong with the EMH and its built-in assertion that above-average performance can be achieved only by taking above-average risks.

Most technical methods for predicting the direction and timing of price movements are based on past price patterns, although some like the January effect or the rule "Sell in May and go away" relate to particular times in the year when markets have historically gone up or down the vast majority of the time.

By now the reader should be getting the impression that there are almost as many market forecasting procedures as there are schemes for predicting horse races, the outcome of football games, or the results of elections. And that impression would be absolutely correct. Moreover, as noted earlier many of them actually work! But very few of these schemes offer a historical perspective and a reasonably solid scientific basis from which to divine the gyrations of the markets on all timescales. For these reasons, let me now present an outline of my own preferred brand of methodology for discerning the pattern of market movements on the various timescales we need for understanding the relationship between stock price movements and human events via the medium of the overall social mood in a population.

I hasten to yet again emphasize a cardinal point: This is *not* a book on stock market methodology. But as stock price movements are our sociometer of choice, we have no choice but to do our best to understand how prices move if we want to use this movement as an indicator by which to anticipate shifts in the mood of society. With this overriding caveat in mind, the reader is quite free to choose whatever price prediction scheme he or she prefers. Here's mine.

NOT-SO-NEW WAVERS

Each year, the Financial Traders Association sponsors a tournament in which traders compete with monitored portfolios in different categories (stocks, options, and others) over a period of several months. In 1984 the Options Division of the tournament was won by Robert Prechter, who increased the value of his portfolio by nearly 450% in 4 months! To gain some perspective on the magnitude of this achievement, the second-highest gain in this category was only 84%, and over 80% of the competitors actually lost money. Prechter described his victory with characteristic modesty when he stated that "There are people who hate the idea that anyone can be successful at predicting the market. I just try to avoid them. But sometimes you walk out on the dusty streets at noon, and there's some guy standing there with his guns."

Given his margin of victory, as well as his previous record as a Wall Street pundit, Prechter had his 15 minutes in the sun during the bull-market price run-up to the Black Monday crash of October 19, 1987. During that time he was busy issuing regular guru-like pronouncements on the state of the market to his avid followers from a redoubt in the wilds of north Georgia. But what exactly is the method that Prechter employs when he predicts the direction of market movements? Interestingly, Prechter's method is not new, but an idea that's been around for 80 years or so and resurrected by him. It now goes under the label the *Elliott Wave Principle (EWP)*. But who was "Elliott"? And how did he come to develop this principle of the way collective human activity like a financial market unfolds?

Ralph N. Elliott was a telegraph operator, a railroad executive, and finally an accountant specializing in restructuring ailing firms. He took up consulting jobs in many parts of the world, and while working in Guatemala in 1927 contracted an amoebic infection that developed into anemia. Having lost part of his savings on Wall Street in the Great Crash of 1929, Elliott moved back to Los Angeles. In frail health and out of work during the Depression, he had plenty of free time to wonder where his money had gone when the market crashed. His analysis convinced him that there is a repetitive pattern that market indexes like the DJIA go through, and that an understanding of these price "waves" would enable the alert investor to see things like October 1929 (and 1987) coming and get out of their way. Of course, to say the Dow moves in a pattern is not really saying much. What we need to know is the *kind* of pattern(s) that the market goes through. And it is just this sort of information that the EWP provides.

As an aside, Elliott was extremely interested in bringing his ideas about the movement of financial markets to the attention of the world, at large, leading to a volume published in 1946 with the touchingly modest title *Nature's Law: The Secret of the Universe*. Parts of it dabble in the Pyramids, Pythagorean number mysticism, and the like. In this sense Elliott was a bit like Isaac Newton, who developed a theory of particle motion that changed the world, but who spent the majority of his life engaged in mysterious alchemical experiments, along with searching for hidden meaning in biblical passages. Anyway, let's see what Elliott came up with by way of explaining—and predicting—the ups and downs of the stock market.

The basic idea of the theory is rather simple to describe. The fundamental pattern upon which the entire Elliott edifice rests is a sequence of up and down price movements constituting a complete cycle of eight waves. Waves 1, 3, and 5 are called *impulse* waves, while waves 2 and 4 are termed *corrective* waves. An entire countertrend movement unfolds in the three lettered waves, a, b, and c. Thus, one complete Elliott cycle consists of eight waves divided into two distinct phases: Numbered phases are in the direction of the main trend, while lettered phases move against the trend. The reader can see a full explanation of this scheme for labeling stock price movements in Appendix A.

It's crucial to understand that these patterns are the direct consequence of basic human psychology, essentially the sum total of the bets traders around the world are placing about the future. We've said before that thinking of an "outside" force causing the pattern already sets you off on the wrong track. Price movements in Elliott's view—and ours—do not follow from the latest Producer Price information, policies of the Federal Reserve, or what the President said in last night's fireside chat; rather, they are the result of the way mass psychology works within a population. Nothing more—or less.

From a theoretical point of view, market behavior and its description by the EWP are every bit as scientific as celestial mechanics, heat diffusion, or the formation of a star. It is governed by definite principles and laws. But they are most definitely not the same type of laws found in the physical sciences, which incidentally is why you'll never understand financial markets by using the tools and methods of physics. A strictly mechanical system will never work to characterize the infinity of possible variations on a single form that the EWP presents. Elliott discovered a basic pattern and all collective behavior of populations of interacting human agents follows one of a multitude of variations upon this fundamental theme. But a huge set of possibilities does not mean absolutely *anything* goes. The market can do what it wants. But it will always do it within the confines of the family of wave patterns discovered by Elliott. These patterns are the "rules of the game." Appendix A contains a description of these rules.

In a very direct sense, the EWP labeling is rather analogous to weather forecasting. The weatherman has information on pressure, temperature, relative humidity, and other data that are processed via known physical laws to generate a pattern of high and low pressure and temperature zones. These patterns are then interpreted against past patterns to render a judgment as to what is the most likely way the pressure and temperature zones will unfold. These possibilities, in turn, give rise to forecasts like "The chance of rain tomorrow is 60%."

The Elliott patterns involve a variety of rules, some of which are strict and must be obeyed, while others are more like rules of thumb but rules nonetheless. So it's natural to wonder whether a computer couldn't just process the price history chart and sort out the various possibilities in which the waves, subwaves, sub-subwaves, etc. would be labeled in a fashion consistent with all the rules. In fact, there are such programs and they do a reasonably good job of identifying the various wave labelings, some even examining databases of past patterns in order to give some labelings a higher likelihood than others. Again, readers interested in obtaining such a program can consult the References section (or the Internet) for pointers to vendors.

Rules and computer programs notwithstanding, it's important to emphasize that mathematics plus computing does not equal magic. No amount of sophisticated science is going to create information that is not already there in the data. The Elliott rules under-determine the wave labeling. So there will almost always be alternative labelings consistent with all the rules. Generally, there will also be good reasons for attaching more weight to some labelings than to others. But not always. To ignore or not ignore is the essence of the *art* of the EWP, not its science.

Speaking of science, we've come to a point in our narrative where it's worth considering the degree to which the social mood theory of human events is "scientific," in any of the senses generally used for that term in polite academic discourse.

SCIENCE FICTION OR SCIENCE FACT?

The protagonist in Isaac Asimov's classic science-fiction trilogy *Foundation* is Hari Seldon, a man who practices what Asimov termed *psychohistory*. Reduced to its basics, this is a procedure not unlike socionomics for using the collective social mood (group psychology) to project future social events. In the 1940s when Asimov was putting together the material for these novels, the very idea of a "science" of history would have been considered sheer ... fiction, at least by most historians and other social thinkers. But that was 60 years ago. And it was also before the rise of complexity science and the age of cheap, powerful computers for simulating complex social processes. These factors have changed dramatically the picture of what we can know about human social activity by *scientific* means. Socionomics takes this entire process to a completely new level. So can we say that socionomics represents a true science of human events of the sort that Hari Seldon imagined? Probably not, at least not quite yet.

The typical path that a domain of inquiry follows to become a full-fledged science involves the following steps:

Observation/Data: Observations of the phenomenon of interest and collection of data summarizing these observations. So, for instance, if the phenomenon of concern is planetary motion, we make many observations by eye, telescope, or otherwise, and then translate these observations into data (often expressed numerically).

Empirical relations: The data are then synthesized into a small number of empirical relationships, such as Kepler's three laws of planetary motion. This is the first stage in the development of a scientific *theory/hypothesis*.

Theory: A theory based upon rules encompassing many phenomena, including the one of concern, is developed within which the empirical relationships are consequences of the rules. So, for example, Newton's inverse-square law of gravitation is a rule that can be expressed mathematically and which has as a *consequence* all three of Kepler's laws. But the theory also tells us about ocean tides, the trajectory of a thrown football, and many other phenomena besides the movement of planets.

Test of hypotheses: The provisional theory or hypothesis makes many predictions about what we should see under various circumstances. These predictions are then tested under controlled conditions in repeatable experiments in a laboratory to either add confirmation to the theory or to suggest modifications of it to better accommodate the results of the experiments.

Presently, socionomics seems to be at the empirical relations stage of development. This book has presented a vast array of evidence in support of a strong correlation between social mood polarity and social events/behaviors/actions on all timescales. But as we saw earlier, a correlation does not necessarily constitute a cause. Only further research will help clarify this matter in the context of social mood and events.

Another crucial part of the puzzle that's currently lacking is a deeper theory for why human brains acting in consort give rise to the herding mentality and group psychology that we have called the social mood. Development of such a theory is central to the elevation of socionomics from a set of empirical observations and correlations to a science. The adventuresome reader will find a tentative research program outlining what I believe is needed to move this entire initiative forward in Appendix B.

To end this brief excursion into the philosophy of science, it's entertaining and instructive to follow the work of David Calderwood and compare Asimov's psychohistory with our theory of the flow of human events. First, the similarities.

Both socionomics and psychohistory

- Are based upon the concept of mass psychology and emotion as the driving force behind collective human events;

- Rely upon the individuals forming the group remaining unaware of the way the group mood works to bias events. Psychohistory does this through secrecy, while socionomics accomplishes the same goal through the misguided belief by the vast majority of people that events are the driver of change rather than feelings, emotions, and beliefs;

- Offer a probabilistic, not deterministic, view of the future.

The differences are perhaps even more interesting than the similarities. They include

- The way psychohistory is used is to direct social change along a line seen as beneficial for all. In Asimov's story, Hari Seldon undertakes a program to prevent the widespread destruction that will follow the decline of the galactic empire. Socionomics, on the other hand, has no ulterior motives. It is simply a tool to be used for understanding and forecasting the world we see around us every day.

- A difference in their foundations. Psychohistory stems from pure mathematical abstractions, while socionomics emerges from the identification of patterns seen through careful observation of real-world data.

- A huge difference in their causal basis. Psychohistory retains the physics-based "event causality" structure that assumes social mood is affected by political, cultural, and economic events ("news") of all sorts. Socionomics, on the other hand, renounces this causal structure, assuming instead what we have called "social causality" by which everyday human interaction gives rise to an *endogenous* social mood and an intrinsic dynamic. That mood, in turn, biases the character and form of all social action.

So it is the reversal of the ingrained notion of event causality that is the essential difference between psychohistory and socionomics. And it is the explication and elaboration of that central difference that has been our leitmotiv in this book. We cannot emphasize this point strongly enough, since no true understanding of human social processes is possible without first renouncing the world view of Newtonian mechanics when you direct attention to the human behavioral realm. In short, finance is not physics—and neither is any other aspect of human activity. It's time to give Newton a rest!

From the Fundamental Working Hypothesis expressed earlier, as well as many of the examples presented throughout the book, I think it's clear to even the most casual reader that on a long-term (decade or more) basis, the world seems to be transiting from a period of an unrelenting positive social mood that dates back at least to the mid-1970s, to what gives every appearance of being a decades-long period of increasingly negative social mood. In our final section we make some carefully chosen forecasts for this period, as well as examine the obvious—and pressing—question: What can be done? Is there anything we can do as a society to mitigate the negative changes this turn in mood will engender? Can an enlightened populace guided by political and business leaders of good faith do anything today to brighten up such a downbeat picture of tomorrow? And what about each of us as individuals? What can we do to help prepare ourselves and our families to survive this storm and come out into sunshine when it finally ends (which it surely will)? These are the issues with which we now conclude our story.

INTO THE ABYSS—AND BACK OUT AGAIN

In the run up to the US presidential election in 2008, Austrian friends continually asked me how I thought it would turn out. Since early 2007 my reply was always the same: an easy win for the Democrats, regardless of their candidate. And so it was. What's more interesting than the outcome, however, is why it was so easy to call this election. We've already seen the reason: In the entire history of American presidential elections there has never been a case when the incumbent or their party survived when the voting public was pessimistic about the future at the time of the election. As Prechter and Goel have shown, sometimes this tendency applies to entire parties when there is no incumbent. By 2007 it was manifestly clear that public frustration with the Republican Party had reached historic levels. They simply had to go. The Democrats could have run "Mickey Mouse" as their candidate and he'd have been a landslide winner.

As I live near the seat of the Austrian government in the center of Vienna, it's of more than passing interest to me to see how the same kind of mood plays out on this side of the Atlantic. Examination of this question is instructive as an exercise for the reader in applying the tools we've developed for polishing the crystal ball just a bit brighter. So let me tell an abridged version of that story to illustrate this point.

The same phenomenon as in the US was at work in the Austrian elections in September 2008, when the leading parties from the 2006 elections were trashed by the voters. In the Austrian case, however, the voting public had several more-or-less equally unappealing candidates to spread their frustration among instead of just one. But as in the US, the public mood in Austria had shifted dramatically to the negative and the total fraction of the vote for the two leading parties saw a 20 percentage-point drop from where it had been just two years earlier. (As an aside, this same negative view of the future is certainly the underlying reason why governments of EU nations were uniformly terrified of calling for a popular referendum in their respective countries on ratification of the controversial Lisbon Treaty.)

What's happening in the case of these elections is an illustration of the far more general principle we've focused on throughout the book: Public frustration (negative social mood) translates into regime change (a discontinuous shift in social behavior). We have chosen the financial market index as a measure of social mood. So when it comes to something like political action in Austria, the appropriate meter stick is the financial market in Vienna, not New York. Figure 6.12 shows the ATX index for the Vienna Stock Exchange on a monthly scale from 1993 to the end of 2008. The chart notes several of the more visible social, political, and economic events in Austria over that period, as well.

This figure has several points of interest. First, we note that not much was happening in Austria until around the middle of 2003. Of course there was the usual smorgasbord of political infighting, governmental posturing, the odd murder, and a financial scandal or two just to keep things lively and the newspaper presses humming. Moreover, from the chart we see that even in this dull period many events took place that were perceived at the time as being positive, such as Austria joining the EU in 1995, along with negative events, like the rightist FPÖ party joining the government in 2000. But when

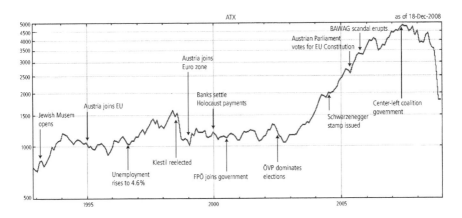

Figure 6.12   The Austrian ATX index, 1993–2008.

the mood started taking off in a resolutely positive direction in 2003, we see mostly upbeat, optimistic types of events such as the issuance of the Arnold Schwarzenegger postage stamp in 2004, along with parliamentary approval of the Lisbon Treaty (without benefit of a popular referendum!) in 2005. Of course, when the collective mood fell off the cliff in October 2007 rather more negative types of events began to dominate, such as the collapse of the center-left government, initiation of a "sex hotline" by the Vienna City Hall to help finance the main public library, and the outcry against the homoerotic-focused Hrdlicka sculpture exhibition to cite a few eclectic examples.

I'll return to the future of Austria in a minute. But for now let me look at the world as a whole, and at what the next few years seem to hold in store. Here are the basic background assumptions to keep in mind for what follows:

- **The long-term global mood has shifted from positive to negative;**

- **This will result in a massive change in every aspect of geopolitical, financial, and social life;**

- **The current problems are psychological, not financial.**

Just in case you've been bamboozled by the run-up in the financial markets since March 2009 and think that the long-term negative social mood is over, think again. Figure 6.13 shows that the entire rally in stocks is attributable only to the unit of measurement, fiat money (banknotes), not *real* money (gold). As the graphic shows, the DJIA topped out at the end of 1999(!) and it's been downhill ever since. So when people ask me about how the "rally" impacts the story told here, my reply is usually, "What rally?"

John Petersen of The Arlington Institute has identified a list of the signs of the "megachange" that's just now begun. Here are a few items from his list:

– The global financial system is collapsing;

– We are at the beginning of the end of oil;

– The global climate system is undergoing huge changes;

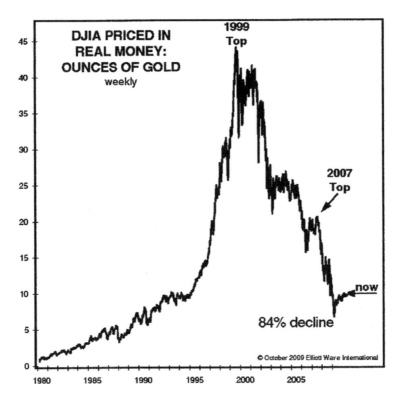

Figure 6.13 The Dow Jones industrial average in ounces of gold, 1980–2009.

– The cost of food is rising dramatically;

– Large solar eruptions are way overdue.

What makes the current situation vastly more dangerous and dramatic than any humankind has previously experienced is that these multiple trends are *converging!* The situation is so complex that no one understands it. But from the perspective of this book we can confidently echo what Michael Jordan said about his retirement: "It's time." Keeping these factors in mind, let's attempt a few general forecasts.

The overriding point underpinning anybody's forecast of economic, financial, political, or social events over the next several years, not just ours, is that the world is now in the opening phase of a decades-long, titanic shift in geopolitical power and individual lifestyles. So a short- to medium-term forecast about anything has to be projected onto the background of this longer-term process; otherwise, a forecast simply cannot be understood. With this caveat in mind, the most straightforward short-term forecast to make is that the global social mood will move sharply toward the negative, giving rise to events in all spheres that could be labeled using everyday words like *discord, fear, anger, somberness, exclusion, separatism*, and the like. Here are a few ways to place your bets taken from different areas of life:

Finance: Many of the corporate entities that have been bailed out by governments worldwide will fail again, along with new ones. The IMF, World Bank, US Federal Reserve, and other financial sources will not bail them out again.

Political: Patriotism and anti-government sentiment will grow into powerful emotional forces. Extremist parties are likely to gain in popularity as old, established parties decline, or even disappear altogether.

Economic: Unemployment will rise to exceed 10% worldwide in the next year or two and ultimately go beyond 25% in many countries.

Social: Professional sports, like soccer and basketball, will suffer difficulties. Attendance and television viewership will decline and player salaries will decrease.

Now what about Austria as a case study at the local, rather than global, level? Unfortunately, I cannot say that the immediate prospects are nearly as lovely as the country. The wave of negative social mood that began in the United States in 2008 has been slowly making its way across the Atlantic in a relentless push eastward. The early stages of its movement through central Europe showed up in the Austrian elections in 2008. But that was just a small preview of coming "attractions." Here are some specific forecasts in order to set the tone for what we might expect to see in the next couple of years in Austria:

Finance: At least one of the three largest banks in Austria will fail, or what's more likely, be forced into an unwanted merger with another bank.

Political: The erosion of the political center that was so pronounced in the 2008 elections will continue, and existing and new extremist parties, left and right, will take even more voters away from the traditional parties.

Economic: Affordable housing will be hard to find. Family members will move in with each other. Homelessness will increase.

Social: People will return to more traditional social values, focusing on friends and family. The rich will be vilified and conspicuous consumption of luxury items will become very "uncool." As an indicator of this process, most restaurants will decline in popularity, and home cooking will become popular (again!).

So there it is. Not much to look forward to actually, other than a more humane way of life. A number of articles in the popular press called 2008 a "lost year," better forgotten as we look to a brighter 2009–2010, which they puzzlingly seem to take as being a return to the never-to-be-seen-again days of 2004–2006. To those indefatigable optimists still clinging to this nostalgic view of the future, I can say only "Dream on"! But not to end on a negative note, we should bear in mind that trends change. And this negative slope we're currently sliding down will too. It's well worth considering how to deal with the negative period so as to be ready when the upswing comes around again.

The first thing to note is that a downturn is not uniformly bad. There is a silver lining in each wave down, and each such wave creatively destroys things that really needed destroying anyway. As we said earlier, who's going to miss the five-dollar cup of coffee at Starbucks or the unlamented Hummer? And this is not to mention junk mail, financial unaccountability on Wall Street, and the Brangelia-inspired "celebutant" media circuses. So there is indeed much to be thankful for in the current turn to the negative.

More seriously, though, we should think about what kind of jobs are likely to be part of growth industries in the future, and try to guide our children (and perhaps ourselves!) into these areas and away from toxic occupations. Table 6.2 presents a short list of both.

To close on a somewhat upbeat note, let's consider what we can do as individuals to cushion the blow from the social "tsunami" we're now facing.

First of all, we can plan for the transition. Think long and hard about how you can provide for yourself and your family. And do not, I repeat, *do not* rely on your government to save you. The situation is serious. And the track record of government action in times that are far outside the norm is not a good one. So it's up to you to save yourself.

Table 6.2 Toxic and growth occupations for the future

Toxic occupations	Growth occupations
Auto worker	Computer programmer
Airline employee	Day-care provider
Financial "services"	Environmental engineer
Supermarket employee	Pharmacist
Fastfood/restaurant worker	Teacher
Internet advertising worker	Data miner
Oil industry worker	Physician's assistant
Media provider	Home health aide

The second thing you can do is begin working with others to sustain yourself. Group action is needed here, as you are unlikely to be able to create a sustainable life for yourself and your family on your own.

Finally, start thinking about the upswing in social mood that will be coming up. This should be the immediate and top priority item for every government, firm, international organization, and individual.

Now for a small word of caution: All the things I've outlined here might not happen! Recall that social mood theory provides a probabilistic forecast, not a certainty. What I've outlined is far more likely than not to unfold. But as with the by-now-mythical "weapons of mass destruction" in Iraq, it's not a "slam dunk." Following John Petersen's sage counsel, in order to keep up with the rapidly-changing circumstances of the coming years I urge you to

- Stay flexible,

- Stay open to new ideas,

and *most importantly,*

- Hang loose!!

APPENDICES

APPENDIX A: ELLIOTT WAVE ANALYSIS

*I*n this short Appendix we give a bit more detail on the Elliott scheme for characterizing stock price movements. This treatment is by no means a full and complete account of how the Elliott wave system works to forecast price movements, but only a bit more detail than was appropriate in the main text for a general-reader volume. The full details are given in the volumes listed in the References.

Elliott found that following completion of the basic pattern shown in Figure A.1, a similar but higher-level pattern begins: another five-wave up pattern followed by another down pattern of three waves correcting the up pattern.

How many cycles, subcycles, and sub-subcycles are there? Elliott identified at least 13 distinct wave degrees, the largest of which he called the Grand Supercycle, waves he thought lasted for decades and went back to the founding of the United States. At the bottom of his hierarchy is the most microscopic Subminuette cycle consisting of waves lasting just a few minutes or less. The chart below shows nine of the degrees identified by Elliott, together with a very logical and easy-to-use numbering scheme developed by Robert Prechter for labeling them.

The basis of this scheme is that we start at the lowest of the nine impulsive degrees, the Subminuette, and label it and the two levels above it with Roman numerals. The three degrees above that are given Arabic numerals,

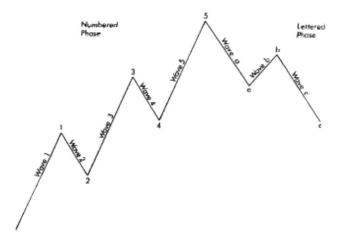

Figure A.1   The basic Elliott wave pattern.

with the next three degrees returning to Arabic. Similar labeling is done for the corrective degrees, which alternate between three lowercase and three uppercase letters. Moreover, the Roman symbols are all lowercase below Minor degree, uppercase above it. Finally, the symbols are distinguished in groups of three by being encased in circles or parentheses or standing alone without enclosures. The essence of the Wave Principle is that moves in the direction of the trend of the wave of next higher degree are five-wave patterns, while moves against the trend of the next larger degree are three-wave moves. This provides the basic information necessary for understanding which direction prices will move next.

The overall Elliott wave pattern is shown in Figure A.1, while Figure A.2 shows the numbering scheme and carries the idea through one entire market cycle. In the second figure we see that each of the numbered and lettered phases is actually a wave itself, but of one degree higher than its component waves. This figure shows explicitly a few impulse waves marked with I, along with their corresponding corrective waves circled and marked with C.

Elliott discovered a number of patterns formed by these elementary cycles. Here are the rules constituting the distilled essence of Elliott's discoveries governing impulsive waves:

- Wave 2 cannot fall below the start of Wave 1.

- Wave 3 cannot be the shortest wave among Waves 1, 3, and 5.

Wave Degree	5s With the Trend (↑ next is Arabic symbols)					3s Against the Trend (↑ next is caps)		
Grand Supercycle	Ⓘ	Ⓘⓘ	Ⓘⓘⓘ	Ⓘⓥ	Ⓥ	ⓐ	ⓑ	ⓒ
Supercycle	(I)	(II)	(III)	(IV)	(V)	(a)	(b)	(c)
Cycle	I	II	III	IV	V	a	b	c
Primary	①	②	③	④	⑤	Ⓐ	Ⓑ	Ⓒ
Intermediate	(1)	(2)	(3)	(4)	(5)	(A)	(B)	(C)
Minor	1	2	3	4	5	A	B	C
Minute	ⓘ	ⓘⓘ	ⓘⓘⓘ	ⓘⓥ	ⓥ	ⓐ	ⓑ	ⓒ
Minuette	(i)	(ii)	(iii)	(iv)	(v)	(a)	(b)	(c)
Subminuette	i	ii	iii	iv	v	a	b	c
	(↓ next is Arabic symbols)					(↓ next is caps)		

Figure A.2 A complete market cycle.

- Wave 4 cannot overlap Wave 1, other than Waves 1, 5, A, or C of a higher degree.

- Waves 2 and 4 should unfold in two different wave forms. If one is "sharp," the other is "flat."

There are three major types of impulsive waves:

1. An *extended* wave: Among Waves 1, 3, and 5, only one should unfold in an extended fashion, meaning the wave is elongated in nature and subwaves are conspicuous to waves of higher degree. The figure below shows an extended Wave 1, Wave 3, and Wave 5.
2. A *diagonal triangle at Wave 5*: This occurs when the momentum of Wave 5 is so weak that the 2nd and 4th subwaves overlap with each other and evolve into a diagonal triangle. Below is also a picture of a diagonal triangle.

213

Possible Extended Waves

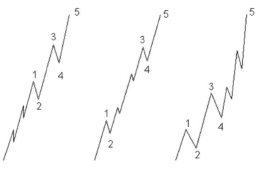

3. A *Wave 5 failure*: This situation arises when Wave 5 is so weak that it cannot surpass the top of Wave 3, causing a double top at the end of the trend. The figure below shows an example of a Wave 5 failure.

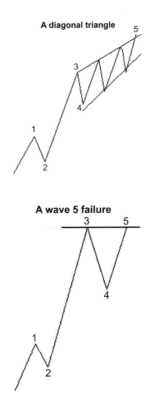

Elliott also identified the personalities of corrective waves, which differ in many important ways from their impulsive counterparts. He found there

Six Corrective Patterns

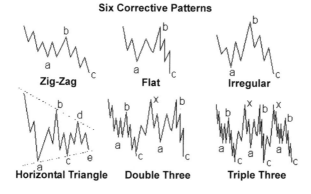

are six basic forms that such waves can take. All six forms are displayed in the diagram above. Each of these patterns has a set of rules of its own. For instance, the Zig-zag is a pattern in a corrective wave satisfying the conditions that it is an A-B-C pattern composed of a 5-3-5 subwave structure. Similarly, the Flat is an A-B-C pattern composed of a 3-3-5 subwave structure, with Waves A and B having the same magnitude.

We won't go into further details of the other patterns here, as the reader has surely already gotten the general idea. These rules are not cast in concrete. But they are good guidelines.

It's of considerable importance to note that the time frame does not enter into the pattern, i.e., the waves may be stretched or compressed without losing the underlying pattern. Elliott discovered not only the fractal character of financial price movements, but also what mathematicians call a "topological" property of these price curves: They look the same if you draw them on a rubber sheet and then pull the sheet in any fashion without tearing it. The only property that's preserved under such operations is the type and relative location of the turning points of the overall wave pattern. The time duration and the magnitude of the price movement can be almost anything, subject to Elliott's rules cited above. This having been said, there do in fact exist rules of thumb governing the likely magnitude of the extended wave and the corrective waves that we'll consider momentarily. And, of course, each wave of higher degree unfolds over a longer timescale than all of its component waves of lesser degree.

The diagram in Figure A.3 shows how the EWP works in practice. Here the various cycles and subcycles are marked for the DJIA over the period 1930 to early 2003. In Figure A.4 we zoom-in on the single Wave V in the preceding figure, and see its fine-grained structure. This graphic illustrates

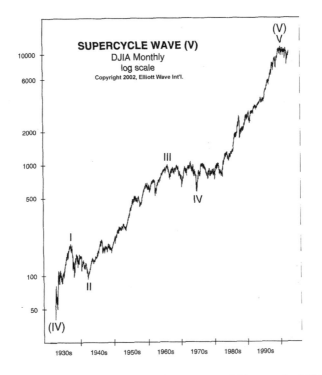

Figure A.3 Wave structure of the DJIA, 1930 to early 2003.

how the Elliott pattern can immediately relate current market action to the main underlying trend. This is what we meant earlier by the remark that the EWP gives a historical perspective on market movements, hence upon fluctuations in overall social mood.

It's clear that the EWP can get a bit complicated at times, especially in corrective waves, as we have to look for patterns within patterns within patterns to nail down just what's going on. The biggest problem is that sometimes it is far from evident if the internal structure is a 3-wave or a 5-wave pattern, which means that we have to develop alternate labelings for both internal structures and look for other ways to assess the relative likelihood of the different candidates. This additional information might be other market indicators like momentum, bullish versus bearish sentiment, trading channels, and of course, just more data as we often have to just wait for the market to unfold and reveal which candidate pattern is the right one.

What about the timing and the magnitude of the waves, up or down? Elliott's work says nothing at all about these matters, but only describes the shapes of the waves and their structures relative to one another. But timing

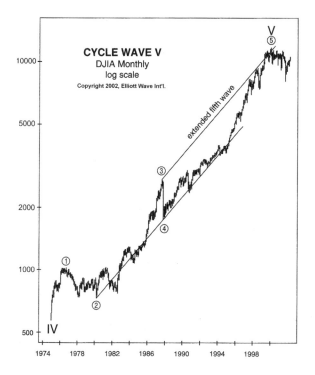

Figure A.4   Wave V of the DJIA, 1975 to early 2003.

and magnitude are crucial, at least for traders. It turns out that the famous Fibonacci number sequence that turns up in the most unlikely places in both natural and human phenomena can help here.

Suppose you start with a bear- and a bull-market price series consisting of a single down wave followed by an up wave. The two of them together constitute one market cycle of two waves. Then take a three-wave bear correction following the bull-market wave, followed by a five-wave upward impulse wave of the new bull market. This phase consists of the three waves in the down phase, followed by five waves up, for a total of eight waves. Continuing this process, we obtain the sequence shown in Figure A.5. The numbers to the right of the figure will be familiar to every math buff. They are the famed Fibonacci sequence, obtained by starting with the two numbers 1 and 1, the rest of the sequence formed by adding the two previous numbers. Using this rule, the first few numbers in the Fibonacci sequence are 1, 1, 2, 3, 5, 8, 13, 21, 34, ...

The so-called "golden ratio" is of special interest for reasons explained in the volumes cited in the References. If we define f_n to be the nth number

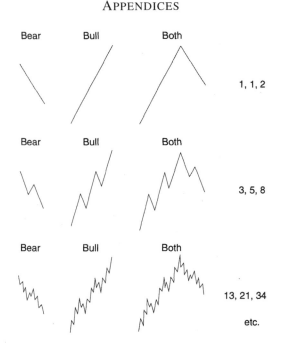

Figure A.5   Financial price waves and the Fibonacci numbers.

in the Fibonacci sequence, then the golden ratio s is defined to be the limit of the ratio f_{n+1}/f_n as n becomes very large (that is, goes to infinity). It turns out that the number $s = (1 + \sqrt{5})/2$, which is approximately $0.618033\ldots$ Note that $1 - s = -1/s$, a relationship of considerable utility in practice. Let's now see what all this number mysticism has to do with Elliott waves.

The simplest way the Fibonacci numbers and the golden ratio appear in the EWP setup is shown in Figure A.6, which displays the familiar five-step impulse wave followed by the corrective three-wave retracement of the

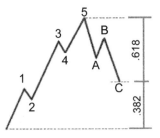

Figure A.6   Elliott waves and the Fibonacci retracement.

218

overall gain. Note carefully that the retracement of the five-wave trend is a Fibonacci ratio $s = 0.618\ldots$ of the total movement in the direction of the trend. Moreover, the full eight-wave move is a fraction $1 - s$ of the peak at the top of wave 3. This is the most common rise-and-fall pattern for the Elliott wave structure, and provides a solid basis for estimating how far a trend will reverse once it tops (or bottoms), as well as estimating the magnitude of the total move in the overall impulse plus correction waves. Some traders believe that a retracement of $1 - s = 0.382$ from a trend move suggests continuation of the trend, while a retracement of length $s = 0.618$ implies a change of trend in the making.

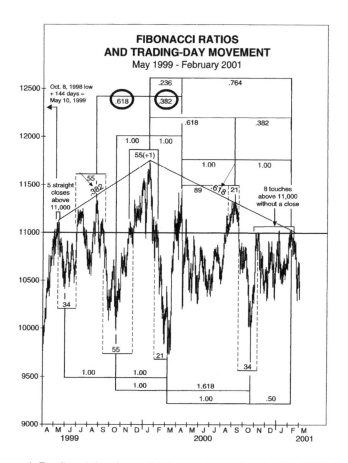

Figure A.7 Special ratios and Fibonacci numbers in the DJIA, May 1999–February 2001 (Copyright 2001, Elliott Wave International).

The above discussion is purely theoretical and abstract. The chart in Figure A.7 above shows the DJIA from mid-1999 through early 2001, which includes the time when the Dow reached a major high on January 14, 2000. We see clearly in this chart how these special ratios and the Fibonacci numbers enter into the timing of peaks and troughs. This diagram also shows some of the ratios *between* the highs and lows over this period of time, which covered exactly a Fibonacci number of months, 21.

We see that the important high on January 14, 2000 subdivided the highs adjacent to it on August 25, 1999 and April 11, 2000 into fractions, $s = 0.618$ and $1 - s = 0.382$, of the whole shown as circled on the chart. The arrows on the figure display the fact that the January 14, 2000 high also separated the time span between the first sustained closing high above 11,000 on May 13, 1999 and the average's last touch of 11,000 on February 13, 2001 into fractions 0.383 and 0.616 of the 21-month period, an almost perfect Fibonacci relationship. Many other Fibonacci ratios and numbers are also given on this chart, which stands in strong testament to the strange

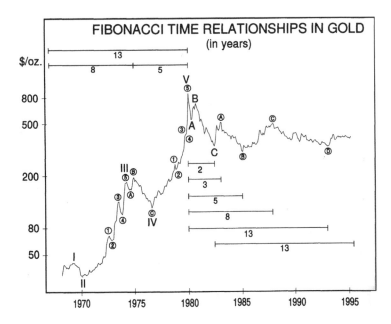

Figure A.8 Fibonacci numbers of months between successive peaks in gold prices, 1970–1995 (Copyright 1995, Elliott Wave International).

power of the Fibonacci numbers to organize the ups and downs of a financial index like the DJIA.

To see the way the Fibonacci numbers enter into time spans of years, not just months, we have Figure A.8 above showing the movement in gold prices over the 25-year period 1970–1995. Again, there is clear evidence of Fibonacci numbers entering into the time spans between different peaks and troughs in the fortunes of the "gold bugs."

These Fibonacci time spans are not just useful in retrospect, but can be used to forecast the future as well. In 1980, Robert Prechter used the Fibonacci time spans of 21 years and 34 years from 1967 to forecast a major bottom for gold on "New Year's Day of 2001 ± one month." Six years later the end of the corrective process in gold occurred at $253.50 on February 16, 2001, from which point gold has advanced nearly 400%.

APPENDIX B: A SOCIONOMICS RESEARCH PROGRAM

The theory of social mood and events ("socionomics", as termed by Robert Prechter) is in its infancy, based as it is upon numerous empirical observations and plausibility arguments for explaining what we see. Many researchers have contributed to the development of the field over the past decade or two. These include Wayne D. Parker, Michael Greene, Ken Olson, John Nofsinger, Kenneth Kim, Hernan Cortez Douglas, Deepak Goel, Peter Kendall, and Alan Hall. To become a bona fide theory of such phenomena, every one of the components of the Central Hypothesis of Socionomics presented in the book must be supported by research efforts and controlled experimental findings. Before stating the research program we have in mind, let's re-state the Central Hypothesis of Socionomics in slightly different form than earlier, just for reference:

Herding Instincts + Interaction → Social Mood → Collective Events

At its deepest level, the research program outlined here aims to explore each aspect of this "dogma," and to understand both the individual elements and the causal arrows linking them. In short, the overall goal is to refute or confirm the flow of information and timing built-in to the Central Hypothesis.

One of the cardinal principles underlying this research effort is that the movements of prices in financial markets, index movements in, say, the Dow Jones Industrial Average (DJIA) or the Standard & Poor's 500 Index (S&P 500), are described on all timescales by patterns originally discovered by Ralph N. Elliott in the 1930s. These so-called Elliott waves are thus fractal in character, looking the same at whatever timescale one makes observations. A crucial aspect of the research proposed here is to provide a scientific basis for this empirical evidence and to test the hypothesis that the Elliott wave patterns for financial time series serve as a kind of "thermometer" for social mood, in general.

Now we look more deeply at the various components of the research program suggested here. Basically, the plan involves four principal components: *Data and Computation* (tool development), *Social Mood analysis, Social Action*, and *Modeling and Exposition*. Each of these components has its own characteristic questions and modes of analysis, as indicated below.

DATA AND COMPUTATION

The basic thrust of this part of this program centers on the gathering and analysis of data generated by both human and non-human (natural) sources, in order to analyze the data for Elliott wave structure. Such a step involves acquisition of data from financial and non-financial, but still human-generated, sources, as well as investigating data time series from various parts of the natural sciences, such as engineering, chemistry, and geophysics. Once these data series are at hand, a computer program may then be used to address the following type of questions:

- Do all living systems produce Elliott wave patterns of the same sort? To study this question we must investigate natural phenomena unrelated to human psychology to see if they differ in any basic way from the data gathered on processes in which human decision-making plays a central role. Statistical tests need to be performed to identify and significant differences between the two situations.

- What are the statistical dimensions of idealized (i.e., artificially-generated by computer) Elliott wave patterns and how do they compare with those of real stock-market data?

- Starting with a list of Fibonacci price and time relationships in the DJIA for the period 1932 to the present, do these ratios appear significantly more frequently in market data than could be expected by random occurrence?

- What is the best way to computerize the Elliott wave analysis? For instance, is the EWAVES program developed at Elliott Wave International, Inc. better at identifying the wave patterns than competing commercial programs such as the *Elliott Wave Analyzer* or *ELWAVE*?

- Under what circumstances do Elliott wave patterns appear in data obtained from an artificial stock market, such as the one created at the Santa Fe Institute? In other words, what types of restrictions on market structure(s) and/or trader's strategy sets lead EW patterns to appear—or not appear?

SOCIAL MOOD

- What is "social mood"? What type of "sociometers" can one use to measure it? How does social mood emerge out of the "herding" behavior of individuals?

- How can we develop a database of sociometers to measure the shifting patterns in popular culture, such as sports ticket sales, popularity of types of music, and fashion styles? How can we test whether these sorts of social behaviors conform to the Elliott Wave Principle?

- What is the correlation between stock market trends and trends using other types of sociometers?

- What characteristics and patterns portend and/or trigger a change in social mood?

- Can we verify or refute the hypothesis that the majority of financial forecasts are lagged extrapolations of the past, i.e., they produce a picture of the market not as it will be in the future, but as it was in the past?

- What type(s) of network structures (i.e. interaction patterns among individuals) give rise to a discontinuous shift in the social mood?

- Can we verify or refute the hypothesis that the vast majority of economic/financial forecasts never successfully predict change?

- What is the role of the media in generating the social mood?<?TeX ?>

- Can the social mood be manipulated? How and on what scale?

SOCIAL BEHAVIOR

- In what way, if any, does the Central Hypothesis of Socionomics matches up with the Central Dogma of Molecular Biology, which states that $DNA \rightarrow RNA \rightarrow Protein$?

- How can we investigate whether stock market trends reliably follow or lag "predicted" types of social actions and behaviors?

- Does the empirical data support the socionomic hypothesis that the intensity of social actions and behaviors like high unemployment in a society corresponds to the degree and label of the Elliott wave in force?

MODELING

- How should we go about creating a "bottom-up" model of financial markets, in which the Elliott wave pattern of price changes is a consequence of the model, not simply an observed epiphenomenon?

- What criteria and methods should we use for evaluating the effectiveness of an analytical or computational model that predicts changes in social behavior?

FUNDING

- The Socionomics Foundation (www.socionomics.org) has grants available for socionomic research.

REFERENCES

GENERAL REFERENCES

Socionomics has sprung from the deep insights of Robert Prechter, Jr., who has pursued extensively what we have called here "The Central Hypothesis of Socionomics." It is impossible to truly appreciate the insight and understanding of how causation works in the social domain without reading the following seminal works.

[HSB]: Prechter, R. *The Wave Principle of Human Social Behavior and the New Science of Socionomics.* (New Classics Library, Gainesville, GA 1999.)

[PSS]: Prechter, R. *Pioneering Studies in Socionomics.* (New Classics Library, Gainesville, GA 2003.)

During the course of the book, many citations are given to two monthly publications from Elliott Wave International, the *Elliott Wave Theorist*, and the *Elliott Wave Financial Forecast*. For brevity and convenience, these two newsletters will be abbreviated in this section as *EWT* and *EWFF*, respectively.

REFERENCES

OVERVIEW

REVERSE PSYCHOLOGY

The full statement by Keynes pointing out the role of human emotions and feelings in economic decisions is

> Even apart from the instability due to speculation, there is the instability due to the characteristic of human nature that a large proportion of our positive activities depend on spontaneous optimism rather than mathematical expectations, whether moral or hedonistic or economic. Most, probably, of our decisions to do something positive, the full consequences of which will be drawn out over many days to come, can only be taken as the result of animal spirits—a spontaneous urge to action rather than inaction, and not as the outcome of a weighted average of quantitative benefits multiplied by quantitative probabilities.

A paraphrased variant that cuts right to the heart of our argument in this book is "The markets are moved by animal spirits, and not by reason."

Interestingly, while Keynes set high store by this emotional component of economic decision-making, and even emphasized it in his work, the economic profession settled on what is probably the lesser aspect of Keynes' message, that governments should intervene in the economy so as to take up the slack when economic cycles starting pointing downward. Strange! Anyway, the whole story is given in Keynes' magisterial work

Keynes, J. M. *The General Theory of Employment, Interest, and Money.* (Macmillan, London 1936.)

Recently, the economic profession rediscovered human psychology, probably out of necessity following events since 2007. This sea change is exemplified by the following volume by a Nobel prizewinner and a likely future Nobel winner, which is virtually a paean to Keynes' original idea, now dressed up in modern clothing:

Akerlof, G. and R. Shiller. *Animal Spirits.* (Princeton University Press, Princeton, NJ 2009.)

REFERENCES

A very perceptive and thought-provoking volume arguing why Keynes remains the world's preeminent economist more than 60 years after his death is

Skidelsky, R. *Keynes: The Return of the Master.* (Allen Lane, London 2009.)

The article in which the social mood is attributed to the human impulse to herd is

Prechter, R. R. and W. D. Parker. "The Financial/Economic Dichotomy in Social Behavioral Dynamics: The Socionomic Perspective," *Journal of Behavioral Finance*, 8, November 2007, 84–108.

The herding instinct in human societies is an old topic going back at least a hundred years. One of the first studies on this theme was presented in the volume

Le Bon, G. *The Crowd: A Study of the Popular Mind.* (Unwin, London 1896, Dover Reprint 2002.)

Two more absolutely classic volumes on the theme, published a century or more ago, are

McDougall, W. *The Group Mind: A Sketch of the Principles of Collective Psychology with Some Attempt to Apply Them to the Interpretation of National Life and Character.* (G. P. Putnam, New York, NY 1920.)

Mackay, C. *Extraordinary Popular Delusions and the Madness of Crowds.* (Richard Bentley, London 1841.)

There is a large and growing literature supporting the argument that stock market returns serve as a leading indicator of macroeconomic activity. A good overview of this issue is found in the articles

Comincioli, B. "The Stock Market as a Leading Indicator: An Application of Granger Causality." www.econ.ilstu.edu/ uauje/PDF's/issue1996/Granger_Causality.pdf

For a fascinating account of the more general issue of how changes in prices, in general, not just in stocks, affects the flow of human events, see

Fischer, D. H. *The Great Wave.* (Oxford University Press, Oxford 1996.)

Many other studies of this type are available on the Internet and support the idea that what the financial markets are doing today is what the macroeconomy will be doing tomorrow. Our argument in this book is to simply extend this notion from macroeconomic trends to social trends of *all* types.

CHAPTER 1: THE *UN*CONVENTIONAL WISDOM

REACHING FOR THE SKY

The definitive source for everything you might want to know about skyscrapers is the Internet site www.skyscraperpage.com, from which the graphic in Figure 1.1 has been adapted.

Interesting articles describing the relationship between skyscrapers and the economy are

Pesek, W. "The Higher the Top, the Further the Economy Drops." *Bloomberg News*, December 12, 2006.
Gilbert, M. "London's New Skyline May Augur Economic Slowdown." *Bloomberg News*, June 18, 2004.

For an account of the "Debacle Dubai" leading to the renaming of Burj Dubai as Burj Khalifa, see

Murphy, B. "Dubai Mega-Tower 'Last Hurrah' to Age of Excess," Associated Press, December 2, 2009.

THE ART OF THE POSSIBLE

The study of the impact of social mood on the outcome of US presidential elections follows the account given in

Prechter, R. "Ask Not What Your Candidate Can Do for the Stock Market; Ask What the Stock Market Can Do for Your Candidate." *Market Technician*, 61, June 2008, 10–11.

Presidential popularity is a fickle thing, as evidenced by the following article

Milbank, D. "Obama vs. the Liberals: Pass the Tea to the Left." *Washington Post*, December 17, 2009.

FROM GLOBALIZATION TO LOCALIZATION

The phenomenon of globalization—the flattening of the world, to paraphrase jounalist Thomas Friedman's popular book *The World is Flat*, has been extolled, debated, reviled, analyzed, condemned, celebrated, and considered from just about every angle imaginable, so that it's simply not possible to give a coherent reference or two for all such aspects of it. We content ourselves here with listing just a couple that were instrumental in our treatment:

Steger, M. *Globalization: A Very Short Introduction.* (Oxford University Press, Oxford 2003.)

Saul, J. *The Collapse of Globalism: And the Rebirth of Nationalism.* (Overlook Books, New York, NY 2005.)

CHAPTER 2: WHY "STUFF" HAPPENS

THE START OF SOMETHING BIG

Full accounts of the Enron scandal can be found in many places. The one we like best because it centers on the role social mood played in the whole sordid affair is found in

Prechter, R. "Assumption of Event Causality Versus Socionomic Insight: Enron Example." *EWT*, June 2002.

GETTING INTO THE MOOD

Dominique Moïsi's book on the mood of countries and the way that affects their interactions with other nations is truly a *tour de force*, going far toward explaining today's headlines as a kind of "clash of emotions."

> Moïsi, D. *The Geopolitics of Emotion*. (The Bodley Head, London 2009.)

MADNESS, MEMES, AND MOODS

The madness and manias of crowds has been well chronicled. One of the very best treatments is *Crowds and Power* by Bulgarian novelist Elias Canetti, an investigation that certainly contributed to his receiving the Nobel prize for Literature in 1981. In this detailed work, Canetti investigates how crowd behavior arises in domains as disparate as football hooligans to religious congregations.

> Canetti, E. *Crowds and Power*. (Claasen, Hamburg 1960, English Edition: Victor Gollanz, London 1962.)

How group behavior arises from the microactivities of individuals is explored in the easy-too-read introductory volume by Nobel laureate Thomas Schelling in

> Schelling, T. *Micromotives and Macrobehavior*. (W. W. Norton, New York, NY 1978.)

The concept of memes and thought contagion as analogous to contagion by biological agents like viruses is discussed in detail in the volumes

> Dawkins, R. *The Selfish Gene*. (Oxford University Press, Oxford 1976.)
> Lynch, A. *Thought Contagion*. (Basic Books, New York, NY 1996.)

Richard Roll's insightful arguments as to how CEO overconfidence has undermined numerous corporate takeovers is given in

> Roll, R. "The Hubris Hypothesis of Corporate Takeovers," part I. *J. Business*, 59, No. 2, 1986, 199–200.

MOODS AND MEMETICS IN ACTION

The notion that a constellation of opinions by a group suitably processed into a kind of "group mind" view of a question is the essence of the "wisdom of crowds" concept. In general, it works only when each person in the crowd makes their choice *independently* of any other member of the group—exactly the opposite effect of what we see in mass psychology and mass psychosis. Nevertheless, the wisdom of crowds is a powerful concept when the underlying assumption of independence can be justified. For a popular account with many intriguing and entertaining examples, see the volume

Surowiecki, J. *The Wisdom of Crowds*. (Doubleday, New York, NY 2004.)

THE CONNECTIONS THAT COUNT

Further amplification of the text discussion on small worlds, tipping points, and networks is found in the following popular references and compendium volumes:

Blass, T. *The Man Who Shocked the World*. (Basic Books, New York, NY 2004.)
Watts, D. *Small Worlds*. (Princeton University Press, Princeton, NJ 1999.)
Buchanan, M. *Nexus*. (W. W. Norton, New York, NY 2002.)
Barabási, A.-L. *Linked*. (Perseus Books, Cambridge, MA 2002.)
Gladwell, M. *The Tipping Point*. (Little, Brown, London 2000.)
Newman, M., A.-L. Barabasi, and D. Watts, eds. *The Structure and Dynamics of Networks*. (Princeton University Press, Princeton, NJ 2006.)

Popular computer columnist John Dvorak points out that a better term than *tipping point*, at least for technological products looking to take off, would be *escape velocity*, which conveys a less passive image. John Hiler's article on blogging and the tipping point can be found at

Hiler, J. "The Tipping Blog", www.microcontentnews.com/articles/tipping-blog.htm, March 12, 2002.

MOOD METERS

By far the most extensive study of various types of sociometers is the invesstigation by Robert Prechter in

Prechter, R. "Sociometrics—Applying Socionomic Causality to Social Forecasting." *EWT*, September 2004.

In this fascinating article, Prechter distinguishes between leading sociometers, like the stock market index, and lagging sociometers, like media accounts. The chart on the next page shows the relationship between the two.

The study of social mood in Singapore and its measurement both by the local financial index and the image wall procedure described in the text is from

Casti, J. and L. Ilmola, "Looking for Weak Signals of Change in Trends," in *Thinking About the Future*, E. Tan Hong and H. Tiang Boon, eds. (National Security Coordination Secretariat, Singapore, 2008.)

A PREEMPTIVE STRIKE ON EVENT CAUSALITY

Daniel Gilbert's story in the text about the happiness level of people winning the lottery versus those who become paralyzed is told in detail in his very thought-provoking and entertaining book

Gilbert, D. *Stumbling on Happiness*. (Alfred Knopf, New York, NY 2006.)

REFERENCES

**The Relative Timing of Socionomic Actions
Provides a Basis for Social Prediction**

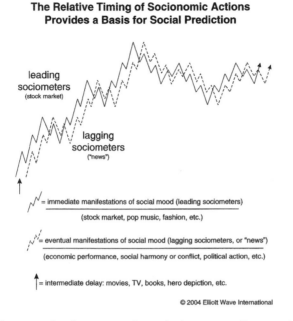

The Kennedy assassination example and charts are discussed along with another example of the period when the US government shut down in 1995 in

P. Kendall, ed. *Prechter's Perspective*. (New Classics Library, Gainesville, GA 2004.)

The study by Mark Jickling for the US Congress of the effect on the stock market of extreme events is reported in

Jickling, M. "The Stock Market's Response to Dramatic Historical Events." *CRS Report for Congress*, Washington, DC, Order Code RS21011, September 14, 2001.

As noted in the text, there has been a sharp revival of interest in Hyman Minsky's ideas in the aftermath of the housing bubble and the ensuing financial crisis. As a result, there are many accounts of this work on the Internet, as well as in Minsky's own books that are now coming back into print. Here are two places to start.

Wilson, S. "Hyman Minsky: Why Is the Economist Suddenly Popular?" *Money Week Magazine*, April 13, 2007.

Minsky, H. *Stabilizing an Unstable Economy*. (McGraw-Hill, New York, NY 2008.)

CHAPTER 3: WHY HITS HAPPEN

PAINT IT BLACK

A very informative study of cycles in women's fashion and social change and how they impact each other is given in the paper

Barber, N. "Women's Dress Fashions as a Function of Reproductive Strategy." *Sex Roles*, 40, 1999, 459–471.

For the curious, The Hot Waitress Index is described more fully in

Lindgren, H. "Hot Waitress Economic Index." *New York Magazine*, August 2, 2009.

The relationship between social mood and the popularity of different car colors is explored in

Galasiewski, M. "Social Mood and Automobile Car Colors." *Socionomics Institute Report*, Gainesville, GA, December 2008.

The sad state of the art auction market is seen almost daily in the general press. A couple of headlines just one month apart from Bloomberg articles about the two greatest auction houses gives the general thrust:

Reyburn, S. "Sotheby's, Christie's Contemporary-Art Sales total Drops 75%." Bloomberg News, December 16, 2009.

The study on the curves of Playmates and their correlation with social and economic times is reported in

Pettijohn, T. and B. J. Jungeberg. "Playboy Playmate Curves: Changes in Facial and Body Feature Preferences Across Social and Economic Conditions." *Personality and Social Psychology Bulletin*, 30(9), 2004, 1186–1197.

HOLLYWOOD HITS AND BROADWAY NIGHTS

More details on the film types and their connection with social mood can be found in *HSB*, as well in the January 2007 issue of the *EWT*.

THE SOUND GOES ON

The material on music and mood follows that given in *PSS* in the article, "Popular Culture and the Stock Market."

THE SPORTING LIFE

For a socionomic account of the travails of baseball as it relates to over-all social mood, the January and October 2008 issues of the *EWFF* are recommended.

POP GOES THE CULTURE

The full list of the 10 things we won't miss about the pre-crisis world from Spike.com was posted at the magazine's web site on August 21, 2009.

The "reverse logic" article on pop culture and the stock market mentioned in the text is

Krantz, M. "Could the Pop-Culture Mood Mirror Stock Market Swings?" *USA Today*, December 19, 2009.

CHAPTER 4: WHY WARS, ECONOMIC CYCLES, AND POLITICAL CRISES HAPPEN

A TIDE IN THE AFFAIRS OF NATIONS

The little vignette between Professors Fate and Chance is adapted from the volume

Casti, J. *Searching for Certainty.* (Morrow, New York, NY 1991.)

Dewey's charts on the cycles of war, together with similar charts on cycles in every aspect of life, are found in the entertaining and illuminating volume

Dewey, E. with O. Mandino. *Cycles.* (Hawthorn Books, New York, NY 1971.)

The relationship between social mood and violence is developed in

Galasiewski, M. "Socionomics and the Sudden Wave of Violence." *EWT*, July 25, 2006.

THE BUSINESS OF BUSINESS

A much more detailed discussion of antitrust suits and their correlation with social mood is presented in *PSS*. The detailed study of this same phenomenon by John Nofsinger and Kenneth Kim is given in

Nofsinger, J. and K. Kim. *Corporate Governance.* (Prentice-Hall, New York, NY 2003.)

THROW THE BUMS OUT

Ominously for the Obama Administration, an August 30, 2009 survey reported by *Rasmussen Reports* states that 57% of the US population would like to replace the entire Congress, while another 18% aren't sure how they would vote on that question if given a chance. So a whopping three-fourths of the US population is deeply unsatisfied with the "bums" in office and would like to throw them out.

CHAPTER 5: WHY GREAT POWERS COME AND GO

The idea that history has a pattern that is not random underpins the very notion that there is a topic here to study. But historians try to tease out this pattern in a bewildering variety of ways, ranging from straightforward narrative accounts to very detailed mathematical and computational models more like what one finds in physics than in the social domain. A very lucid and readable account of how historians try to understand the past and its implications for the future is given in

Gaddis, J. L. *The Landscape of History*. (Oxford University Press, New York, NY 2002.)

More on the mathematical side is the following volume that addresses the ups and downs of nation-states:

Turchin, P. *Historical Dynamics*. (Princeton University Press, Princton, NJ 2003.)

The affairs of humans and the climate have always been strongly inter-twined, and never more so than today. For a sampling of work touching on all aspects of this symbiosis, see

Maunder, W. *The Uncertainty Business: Risks and Opportunities in Weather and Climate*. (Methuen, London 1986.)
Winkless, N. and I. Browning. *Climate and the Affairs of Men*. (Harper's Magazine Press, New York, NY 1975.)
Ladurie, E. *Times of Feast, Times of Famine: A History of Climate Since the Year 1000*. (Doubleday, New York, NY 1971.)

Two volumes providing short, but very incisive, insights into the great power struggle the world will navigate over the next couple of decades are

Joffe, J. *The Future of the Great Powers*. (Phoenix, London 1998.)
Kagan, R. *The Return of History and the End of Dreams*. (Atlantic Books, New York, NY 2008.)

THE LONG WAVE GOODBYE—AND HELLO

The long waves in economic behavior were first identified by Kondratiev, although exactly *how long* the waves are is still hotly debated. Kondratiev's original work is in Russian in 1925 and first came to the attention of the

237

western academic community in a German translation the following year. The full English version can be found in

Kondratiev, N. "The Long Waves in Economic Life," *Review of Economic Statistics,* 2 (4), 1979, 519–562.

More recent accounts of where we seem to stand with respect to these waves are

Alexander, M. *The Kondratiev Cycle.* (Writers Club Press, San Jose, CA 2002.)

Goldstein, J. *Long Cycles.* (Yale University Press, New Haven, CT 1988.)

WAR AND PIECES

A deeper discussion of the outbreak of warfare and its correlation with social mood is given by Prechter in *HSB*.

CYCLING THROUGH HISTORY

Spengler has traditionally received bad press for being what some see as a kind of "proto Nazi," mostly I think because his writings call for a return to Authority and have a decidedly racist tone. But it should be said in Spengler's defense that he never joined the National Socialist Party, and in fact regarded them as more interested in patriotic slogans than in serious political action to help Germany. An excellent introduction to his life is available in

Stimely, K. "Oswald Spengler: An Introduction to His Life and Ideas." *Journal for Historical Review*, 17(2), March/April 1998, p. 2ff.

A nice summary of Spengler's work—pro and con—is given in

Oliver, R. *Journal of Historical Review*, 17(2), March/April 1998, p. 10ff.

For Spengler's own words, see

Spengler, O. *The Decline of the West*. (Oxford University Press, New York, NY 1932, Abridged Edition.)

By now, so many scholars have written so much about Toynbee that it's not necessary to review that literature here. For our purposes, what is important is that Toynbee's approach to history was very similar to that of Spengler, namely, to look at history as the rise and fall of civilizations rather than of nation-states. Of course, the font from which all this secondary literature flows is Toynbee's own *magnum opus*,

Toynbee, A. *A Study of History, Vols. I–X*, (Oxford University Press, Oxford 1934–1961.)

An interesting critique of Toynbee's work by 30 scholars from across the spectrum of historians is

Ashley Montagu, M. ed. *Toynbee and History*. (Porter Sargent Publisher Co., Boston, MA 1956.)

Paul Kennedy's argument about how great powers always overextend themselves to become not-so-great powers is given in painstaking detail in his best-selling volume

Kennedy, P. *The Rise and Fall of the Great Powers*. (Random House, New York, NY 1987.)

To create and maintain an empire is like making sausage in that it's not a pretty process. The implications of Kennedy's case for imperial overstretch of the current American "empire" is considered in the article

"Imperial Overstretch?" *The Economist*, 27 June 2002.

A similar discussion is given in

Isenberg, D. "The Costs of Empire." *Asia Times*, February 13, 2004.

REFERENCES

THE DECLINE AND FALL OF GLOBALIZATION

The schism in Belgium is nicely described in the piece

> Caldwell, C. "Belgium Waffles." The *Weekly Standard,* 15(14), December 21, 2009.

Amid all the hoopla and gushing words of support and approval in the 1970s and 1980s for the amazing idea that nation-states should cede their sovereignty to multinational corporations *for the public good*, a number of more thoughtful folks have recently presented cogent arguments for why this wooly-headed notion is doomed to fail. One of the best arguments, although by no means the only one, is that summarized by John Ralston Saul in the text. But it is only a summary and cannot do justice to the real thing. That "real thing" can be found in

> Saul, J. *The Collapse of Globalism: And the Rebirth of Nationalism.* (Overlook Books, New York, NY 2005.)
> Saul, J. "The End of Globalism." *Australian Financial Review*, February 26, 2004.

The progressively chaotic nature of the global geopolitical system is adroitly described using the physical concept of entropy in

> Schweller, R. "Ennui Becomes Us." *National Interest*, December 16, 2009.

BRICS OR BRIC-A-BRAC?

An argument for why China may not lead the world anytime soon is presented in

> Kurlantzick, J. "Dazzled by Asia." *Boston.com*, February 7, 2010.

The socionomic study of Russia referred to in the text is

> Hall, A. "Sizing Up a Superpower: A Socionomic Study of Russia." *Global Market Perspective Special Report*, Elliott Wave International, Gainesville, GA, November 14, 2007.

Another interesting perspective on the future of Russia is found in

Kuchins, A. *Alternative Futures for Russia to 2017*. Center for Strateguic and International Studies, (Washington, DC, November 2007.)

The quote from McFaul and Stoner-Weiss is from the article

McFaul, M. and K. Stoner-Weiss. "The Myth of the Authoritarian Model." *Foreign Affairs*, 87(1), January–February 2008, 68–84.

WHAT ABOUT AMERICA?

The reference for the *Slate* magazine article on "Apocalypse America" is

Levin, J. and C. Wilson. "How is America Going to End?" *Slate*, http://www.slate.com/id/2223285/

The report *Global Trends 2025: A Transformed World*, published by the US National Intelligence Council, is available on the Internet at the NIC site.

THE REALLY LONG VIEW

The thousand-year wave discussed in the text is discussed further in *HSB*, as well as in the July 2004 issue of *EWT*.

CHAPTER 6: GETTING IT TOGETHER

POLISHING THE CRYSTAL BALL

Peak Oil and its social consequences are a theme that's been examined by many over the past few years. Here are a couple of general references that explore several sides of this vexing question. The subtitles of these volumes give a good account of the general drift of these "pro Peak Oil" books. The "no Peak Oil" side is well represented too — but not nearly as much fun to read as the catastrophists!

Deffeyes, K. *Hubbert's Peak: The Impending World Oil Shortage.* (Princeton University Press, Princeton, NJ 2001.)

Strahan, D. *The Last Oil Shock: A Survival Guide to the Imminent Extinction of Petroleum Man.* (John Murray, London 2007.)

Goodstein, D. *Out of Gas: The End of the Age of Oil.* (W.W. Norton, New York, NY 2004.)

Kunstler, J. *The Long Emergency: Surviving the Converging Catastrophes of the Twenty-First Century.* (Atlantic Monthly Press, New York, NY 2005.)

LIFE GOES ON

The material on prediction, explanation et al. in this and the next section is adapted from

Casti, J. *Searching for Certainty.* (Morrow, New York, NY 1991.)

THE ANTICIPATION MACHINE

The implications of an increasingly negative social mood given in this section are simply "reverse generalizations"; that is, they are specific events representative of a broader class of social phenomena. The specific events attract attention but should not be taken too seriouisly; the broader classes of social events of which the specific events are examples should be taken very seriously indeed. *Something* of the type suggested here is almost certain to happen under the Working Hypothesis that the social mood is turning decidedly negative.

In regard to the discussion of terrorism, the thoughts of Lee Harris expressed in the text are but a tip of a much deeper iceberg. For the full story, see Harris's book

Harris, L. *Civilization and Its Enemies.* (Free Press, New York, NY 2004.)

THE END OF A TREND

The discussion of how to read the headlines from a socionomist's perspective, along with the story of the Mei/Moses Fine Art Index, is considered in much greater detail in

Ross, D. "Reading the Headlines with a Trendy Eye." *Progress in Socionomics*, 1, Winter 2004/2005, 9–12.

HOW NOW, MR. DOW?

Technical analysis of the stock market is discussed in detail in Chapter 4 of

Casti, J. *Searching for Certainty*. (Morrow, New York, NY 1991.)

See also

Mandelbrot, B. and R. Hudson. *The (Mis)behavior of Markets*. (Basic Books, New York, NY 2004.)

Malkiel, B. *A Random Walk Down Wall Street*. (W.W. Norton, NY 1990.)

For much greater technical detail, see

Aronson, D. *Evidence-Based Technical Analysis*. (Wiley, New York, NY 2007.)

Bauer, R. and J. Dahlquist. *Technical Market Indicators*. (Wiley, New York, NY 1999.)

NOT-SO-NEW WAVERS

The definitive work on Elliott Wave analysis is the classic volume

Frost, R. and R. Prechter. *Elliott Wave Principle*. (New Classics Library, Gainesville, GA 1978.)

There are several computer programs available to provide Elliott wave labeling for financial data. None of them seem to do as good a job as a competent "Elliottician". One of the better ones, though, is *ELWAVES*, which is available at the web site www.prognosis.nl.

SCIENCE FICTION OR SCIENCE FACT?

The analogy between Asimov's classic work on psychohistory and socionomics is explored in detail in the paper

Calderwood, D. "Life Imitating Art or Vice-Versa?" *Progress in Socionomics*, 2, Summer 2005, 6–9.

INTO THE ABYSS—AND BACK OUT AGAIN

An interesting list describing future growth jobs that don't yet exist, like body-parts maker, nano-medic, and space pilot are outlined in the feature article

"The World in 2020", *Engineering and Technology*, January 23–February 5, 2010.

The converging catastrophes mentioned in the text, along with much, much more food for thought about surviving the next few years, are explained in the compact volume

Petersen, J. *A Vision for 2012: Planning for Extraordinary Change.* (Fulcrum Publishing, Golden, CO 2008.)

APPENDIX A

Full details of the Elliott wave principle are available in

Frost, A. J. and R. Prechter. *Elliott Wave Principle.* (New Classics Library, Gainesville, GA 1978–2001).

REFERENCES

Details on the use of Fibonacci numbers to forecast the turn in gold prices as described in the text can be found in Chapter 17 of

Prechter, R. *At the Crest of the Tidal Wave* (New Classics Library, Gainesville, GA 1995.)

Index

Index

Index

Index

Printed in the United States of America